Core Concepts for the
Civil PE Exam Breadth and Water Resources and Environmental Depth Practice Exams and Reference Manual: 80 Morning Civil PE and 40 Water Resources and Environmental Practice Problems

Study More Efficiently

David Gruttadauria P.E.
PECoreConcepts@gmail.com

Thank you for your purchase!

We are available for full support of any of your studying needs.

Email us at PECoreConcepts@gmail.com

Copyright © 2022 by Bova Books LLC. All rights reserved. No part of this publication may be reproduced, stored, or transmitted, in any form or by any means electronic, mechanical, photocopying, recording or otherwise, without the prior written permission of the publisher.

The contents of this book are for educational purposes only and in no way shall be used as reference in the preparation of structural calculations or documents. No liability will be assumed by the author and the information in this book shall not be used in a court of law.

Table of contents

    I.   Core Concepts Reference Guide

<u>Morning Session</u>

1. Project Planning
    A. Quantity Take Off
    B. Cost Estimating
    C. Project Schedules
    D. Activity Identification and Sequencing

2. Means and Methods
    A. Construction Loads
    B. Construction Methods
    C. Temporary Structures

3. Soil Mechanics
    A. Lateral Earth Pressure
    B. Soil Consolidation/Foundation Settlement
    C. Effective Stress
    D. Bearing Capacity
    E. Slope Stability

4. Structural
    A. Dead and Live Loads
    B. Trusses
    C. Bending
    D. Shear
    E. Axial
    F. Deflection
    G. Beams
    H. Columns
    I. Slabs
    J. Footings
    K. Retaining Walls

5. Hydraulics and Environmental
    A. Open Channel Flow and Mannings Equation
    B. Storm Water Collection and Drainage
    C. Storm Characteristics
    D. Runoff Analysis
    E. Detention/Retention Ponds
    F. Pressure Conduits

G. Bernoulli (Conservation of Energy)

6. Transportation/Geometrics
    A. Horizontal Curves
    B. Vertical Curves
    C. Traffic Volume
    D. Vehicle Dynamics

7. Materials
    A. Soils Classification
    B. Soil Properties
    C. Concrete Properties
    D. Steel Properties
    E. Material Test Methods and Specification Conformance
    F. Compaction

8. Site Development
    A. Cut and Fill
    B. Construction Site Layout and Control
    C. Construction Erosion and Sediment Control
    D. Impact of Construction on Adjacent Facilities
    E. Safety

Water Resources and Environmental Depth

I. Analysis and Design
   A. Mass balance
   B. Hydraulic loading
   C. Solids loading (e.g., sediment loading, sludge)
   D. Hydraulic flow measurement

II. Hydraulics–Closed Conduit
   A. Energy and/or continuity equation (e.g., Bernoulli, momentum equation)
   B. Pressure conduit (e.g., single pipe, force mains, Hazen-Williams, Darcy-Weisbach, major and minor losses)
   C. Pump application and analysis, including wet wells, lift stations, and cavitation
   D. Pipe network analysis (e.g., series, parallel, and loop networks)

III. Hydraulics–Open Channel
   A. Open-channel flow
   B. Hydraulic energy dissipation
   C. Stormwater collection and drainage (e.g., culvert, stormwater inlets, gutter flow, street flow, storm sewer pipes)
   D. Sub- and supercritical flow

IV. Hydrology
　　A. Storm characteristics (e.g., storm frequency, rainfall measurement, and distribution)
　　B. Runoff analysis (e.g., Rational and SCS/NRCS methods)
　　C. Hydrograph development and applications, including synthetic hydrographs
　　D. Rainfall intensity, duration, and frequency
　　E. Time of concentration
　　F. Rainfall and stream gauging stations
　　G. Depletions (e.g., evaporation, detention, percolation, and diversions)
　　H. Stormwater management (e.g., detention ponds, retention ponds, infiltration

V. Groundwater and Wells
　　A. Aquifers
　　B. Groundwater flow
　　C. Well analysis–steady state

VI. Wastewater Collection and Treatment
　　A. Wastewater collection systems (e.g., lift stations, sewer networks, infiltration, inflow, smoke testing, maintenance, and odor control)
　　B. Wastewater treatment processes
　　C. Wastewater flow rates
　　D. Preliminary treatment
　　E. Primary treatment
　　F. Secondary treatment (e.g., physical, chemical, and biological processes)
　　G. Nitrification/denitrification
　　H. Phosphorus removal
　　I. Solids treatment, handling, and disposal
　　J. Digestion
　　K. Disinfection
　　L. Advanced treatment (e.g., physical, chemical, and biological processes)

VII. Water Quality
　　A. Stream degradation
　　B. Oxygen dynamics
　　C. Total maximum daily load (TMDL) (e.g., nutrient contamination, DO, load allocation)
　　D. Biological contaminants
　　E. Chemical contaminants, including bioaccumulation

VIII. Drinking Water Distribution and Treatment
　　A. Drinking water distribution systems
　　B. Drinking water treatment processes
　　C. Demands

        D. Storage
        E. Sedimentation
        F. Taste and odor control
        G. Rapid mixing (e.g., coagulation)
        H. Flocculation
        I. Filtration
        J. Disinfection, including disinfection byproducts
        K. Hardness and softening

IX. Engineering Economics Analysis
        A. Economic analysis (e.g., present worth, lifecycle costs, comparison of alternatives)

    II.   Practice Problems
    III.   Solutions
    IV.   Answer Key

CORE CONCEPTS REFERENCE GUIDE

MORNING BREADTH

Project Planning

Quantity Take-off Methods

Quantity take-off methods are a means for estimating the cost of each aspect of a project. A project consists of many activities and materials all of which are accounted for as items of a project. For example, a project may involve the construction of a retaining wall. There are many activities and materials associated to complete this. Some include excavation, formwork, concrete for the wall, reinforcing steel etc. When contract drawings and specifications are developed, all of these items must be identified. All items also must include a quantity associated it's them to indicate the amount or extent of work for the item. These quantities must be defined by a particular unit of measure which must be appropriate for the action or material. Taking excavation as an example of an item, there must be an amount of excavation associated with it. Since excavation involves removing a volume of material, the most appropriate unit is cubic yard or cubic feet. To estimate the cost of the project, each item has a price per unit associated with it. This price is determined by previous similar work and taking into account the specifics of the particular project. Below is an example of the breakdown of some items associated with an example retaining wall project:

Item	Unit	Quantity	Unit Price	Cost of Item
Excavation	Cu. Yard	50	100	5000
Concrete (Including Formwork and labor)	Cu. Yard	25	1000	25000
Reinforcing Steel	Lb.	500	12	6000
Backfill	Cu. Yard	40	50	2000
Drainage Pipe	Linear Ft	30	5	150

Cost Estimating

Engineering Economics is used to determine the best economic course of action when weighing construction options by incorporating the life-span of alternatives and comparing costs at equivalent times. The following chart provides the most common equations;

Converts	Formula
P to F	$(1+i)^n$
F to P	$(1+i)^{-n}$
F to A	$\dfrac{i}{(1+i)^n - 1}$
P to A	$\dfrac{i(1+i)^n}{(1+i)^n - 1}$
A to F	$\dfrac{(1+i)^n - 1}{i}$
A to P	$\dfrac{(1+i)^n - 1}{i(1+i)^n}$

P = Present Value
F = Future Value
i = Interest Rate
n = Years
A = Uniform Series Value

Project Schedules

Project schedules must be set and maintained to ensure it remains on time and on budget. To determine a project schedule, all tasks must be identified and the length of time (durations) for each task must be estimated. These tasks can then be sequenced by determining what the appropriate order of tasks are. Some tasks must be completed before others can begin. These tasks are defined as predecessors. See the example chart below indicating identified tasks, durations, and predecessors:

Task	Duration (Days)	Predecessor
A	2	
B	3	A
C	2	A
D	1	B
E	2	B, C

This information can then be visualized by producing and activity diagram. First begin by drawing tasks. Start with A:

Then determine which tasks have A as a predecessor. Draw these tasks as well with arrows indicating these tasks are connected:

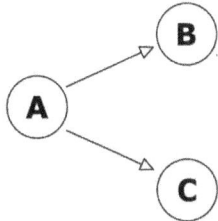

Continue in the same manor with each task. The final chart is as follows:

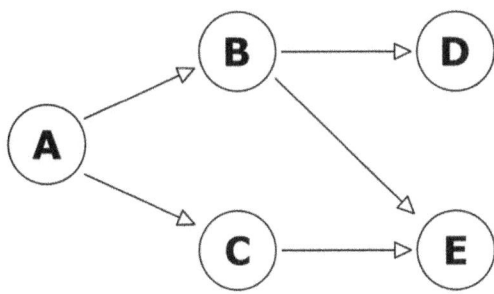

Then you can determine the critical path of the project. The critical path is defined as the sequence of tasks which would yield the shortest amount of time to complete the project. If the duration of any task on the critical path is changed, the duration of the entire project will change. In the example above you can determine the critical path by identifying all paths and the critical one is the longest sum of duration. Therefore, the possible paths are A-B-D, A-B-E, and A-C-E which have total durations of 6, 7, and 6. Therefore the critical path is A-B-E. A change in duration of non-critical tasks will only change the project duration if the change creates a longer path than the existing critical one.

Activity Identification and Sequencing

The appropriate steps in the proper sequence need to be identified to complete a project. This involves understanding all the tasks involved in a specific project type and providing a timeline of events to properly facilitate the successful completion of the project. There are many types of projects and the specifics can vary. For the purposes of the PE Exam, it is important to have a general knowledge of common construction tasks and sequences. Below are some examples of design and construction tasks divided by when they occur in certain project phases:

Pre-Design/Design/Project Award

- Owner initiates project
- Owner hires Architect/Engineer or uses In-House Architect/Engineer
- Contract documents and specifications are developed
- Contractors bid on the project
- Project is awarded

Pre-Construction

- Contractor submittals are reviewed and approved
- Sub-Contractors hired
- Site survey, staking, and layout
- Procurement of materials

Construction

- Traffic Control, water handling, etc. installed if necessary
- Crane set up and positioning
- Temporary earth retaining systems installed if necessary
- Excavation
- Formwork or Erection
- Testing of materials
- Installation of rebar
- Pouring of concrete
- Concrete curing
- Backfill

Post-Construction

- Semi Final/Final Inspections
- Open road to traffic
- Punch-Lists
- As-Built drawings

Means and Methods

Construction Loads

Construction loads are temporary loads, occurring within the duration of a project, imposed on a structure which may be partially or fully complete. This may include materials, personnel, equipment, or temporary structures. The concern for construction loads is to understand the different types of stresses they may impose on members as opposed to the final in-place condition of those members and ensure they are designed to handle these forces.

Materials: Storage of materials is an often overlooked aspect of a project. Rebar, excavated materials, or other building materials need to be stored in an accessible location and will often impose a large additional dead load on the structure.

Temporary Structures: Temporary structures may often be needed to either provide additional support to unstable members or access for personnel to continue the erection process. Temporary structures may also be for the housing of materials or personnel.

Equipment: Equipment is often needed for various construction activities such as welding or painting procedures. The weight and distribution of these loads should be accounted for.

Cranes: Cranes can also be considered equipment however special attention should be given to the sequencing of erection based on the cranes reach.

Members in Temporary Conditions: Along with additional dead load, construction can introduce stresses into members for which they are not designed. Some examples include the erection of a precast member such as a wall panel which may be designed for compression but will see some flexure about its weak axis while it is being picked and placed. Also, the first steel girder in a bridge before it is connected to the others through diaphragms will be unstable and must be temporarily supported. In these conditions, design measures need to be taken even though they are not required for the final condition.

Construction Methods

Steel: Strong and durable material. Steel has the capabilities to be used for long span bridges and high-rise buildings. Steel members are manufactured using either the hot-rolled or cold-formed methods. Steel members are provided in predetermined shapes. Some examples include W-, S-, C, and HSS-shapes. Steel is connected and constructed by the use of bolted or welded connections. The advantages of steel are again the ability to span long distances and the weight of the members compared to the strength is relatively low. Some disadvantages include the high cost, lack of ability to form unique shapes, and tendency of the material to corrode. When steel is exposed to salts, a chemical reaction occurs causing the steel to rust and even loose section properties. To counteract corrosion some preventive measures are paint systems, coating systems such as galvanizing, or weathering steel.

Reinforced Concrete: Concrete is strong in compression but weak in tension. However, it has the ability to bond to reinforcing steel to appropriately resist tension. Reinforced concrete is used in buildings and shorter span bridges or certain components of bridges. Some common applications are foundation elements, bridge decks and parapets, or retaining walls. The advantage of concrete is it can be formed to any shape or aesthetic look with proper formwork and is strong in compression. The disadvantages however are that concrete has a high self-weight, will likely crack, and has a limited span length. The reinforcing in concrete can also corrode and cause pop-outs or spalls.

Precast/Prestressed/Post-Tensioned Concrete: Precast concrete is concrete which is cast somewhere other than its final location, either at a plant or another area on the construction site, and is then stripped from its forming, transported to the site, and erected. Prestressed concrete is precast concrete which has been pre-compressed using steel strands with high elasticity. The strands are tensioned to a design force before the concrete is cast. Then the concrete mix is placed and cured. The strands are then cut at the ends. Since the strands have a high elasticity, they will try to return to their original state. However, since the stands are now bonded to the concrete, there is a compressive force transferred to the concrete. This force will oppose the stresses caused by bending. Post-Tensioned concrete uses the same concept as prestressed concrete. However, the concrete member is cast first and the strands are tensioned through the member using plastic tubes embedded along the length of the member. The tube is then grouted and the strands are cut to transfer the force. Precast Concrete will have the lowest tensile capacity and therefore is used for the lowest spans. Precast is often used for compression members or bridge deck units. Prestressed Concrete is able to span larger distances and is used for floor members. It is common in parking garages as double-tee shapes for floor members or for long span bridges with common shapes such as prestressed bulb tees. Post tensioned concrete is not as common and is used for much larger spans. Precast concrete advantages are the quality of concrete is often better under plant controlled conditions and the construction is much quicker. The disadvantages are there is a higher cost than reinforced concrete due to shipping and erection expenses and the tendons also are susceptible to corrosion.

Wood: Relatively low strength material. Wood is often used in residential applications or for very small span bridges. Wood is extremely cheap and lightweight for erection. In addition to the low strength, wood also will deteriorate due to rot and is highly sensitive to fire damage.

Masonry: Can be reinforced or unreinforced. Masonry is also strong in compression but weak in tension. Only used in small retaining wall applications and some older bridges still are composed of masonry components.

Temporary Structures

Structures which are built for a specific purpose, often to facilitate an aspect of a construction project, and are removed before the conclusion of the project are temporary structures. Some examples include:

- Temporary Buildings: May be used for storage or offices during construction.
- Scaffolding: Temporary elevated platforms which provide access to perform certain tasks.
- Temporary Supports and Shoring: Often forces are introduced during construction which are not the same as the final in-place conditions. In these situations, temporary supports are needed to keep structural members stable until the construction can be complete.
- Temporary earth retaining systems: When excavation is needed, there is often not enough room to safely dig to the required depths. There may be a need to support traffic or adjacent facilities during the excavation. In these cases, temporary earth retaining is required. Some examples include sheet piles, concrete blocks, or trench boxes.
- Formwork: Concrete formwork is used to place concrete to the desired shape and will remain in place until the mix has cured to the desired strength.
- Cofferdams: A wall constructed to prevent the flow of water to a specific area. Can be made of sandbags, sheet piling, or other materials.

Soil Mechanics

Lateral Earth Pressure

Rankine Active Earth Pressure

Resultant Force $R_a = \frac{1}{2} k_a \gamma H^2$ which is applied at a distance of H/3 from the base of the footing

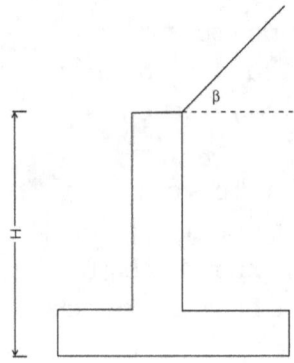

k_a = Active Earth Pressure Coefficient
γ = Density of Soil
H = Height of Retaining Wall from Base of Footing

$$k_a = \cos \beta \left(\frac{\cos \beta - \sqrt{\cos^2 \beta - \cos^2 \phi}}{\cos \beta + \sqrt{\cos^2 \beta - \cos^2 \phi}} \right)$$

β = Angle of Backfill
ϕ = Angle of Internal Friction
If the backfill is horizontal ($\beta = 0$) the equation reduces to:
$k_a = \tan^2(45° - \phi/2)$

Note: Rankine assumes the friction between the soil and wall is zero

Soil Consolidation/Foundation Settlement

Settlement is when the soil supporting the foundation consolidates which causes a decrease in volume and a drop-in elevation. This causes the foundation to no longer be fully supported and will introduce additional stress. There are three phases of settlement:

1. Immediate Settling or Elastic Settling: This settlement occurs immediately after the structure is built. The load from the structure causes instant consolidation of the soil. This is the main component of settlement in sandy soil conditions.

2. Primary Consolidation: A more gradual consolidation which is due to water leaving the voids over time. This is mostly a factor only in clayey soils.

3. Secondary Consolidation: Also occurs at a very gradual rate. This is due to the shifting and readjustment of soil grains. Most often this is the lowest magnitude of consolidation phases.

Effective Stress

The effective stress is the stress at a certain point below grade due to the weight of soil above. It is calculated as the density times the height of each level. However, If the water table is present, the density is reduced by that of water:

Effective Stress = $\Sigma H \gamma$ in dry soil and $\Sigma H(\gamma_s - \gamma_w)$ in saturated soils, where γ_w = 62.4 pcf

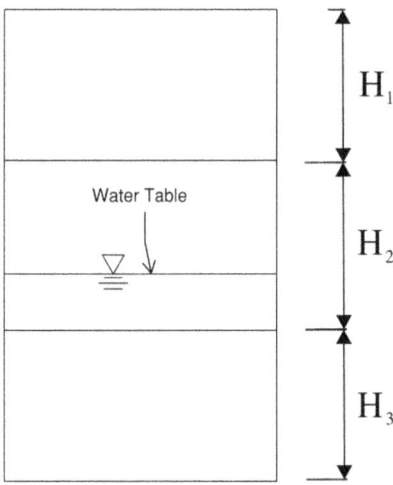

Bearing Capacity

For shallow foundations, the soil below must be suitable to support the load transferred through the footing. Different types of soils have different bearing capacities. Sand is often a good foundation material. Sand undergoes some small immediate settlement and then stabilizes since it drains quickly. Clay generally is poor in bearing capacity. Clays do not drain quickly and will retain water for longer periods of time leading to long-term settlements. Most soils in reality are some combination of sands, clays, and silts which will behave somewhere in-between sand and clay. Exceeding the allowable bearing capacity of a soil will cause shear failure or excessive settlements. Bearing capacity is determined using the Terzaghi-Meyerhof equation:

$$q_{ult} = \frac{1}{2}\gamma B N_\gamma S_\gamma + c N_c S_c + (p_q + \gamma D_f)N_q$$

q_{ult} = Ultimate Bearing Capacity
γ = Soil Density
B = Width of Footing
c = Cohesion of Soil
N_γ = Density Bearing Capacity Factor
N_c = Cohesion Bearing Capacity Factor
N_q = Surcharge Bearing Capacity Factor
p_q = Surcharge Pressure
D_f = Depth from top of Soil to Bottom of Footing
S_γ = Density Shape Factor
S_c = Cohesion Shape Factor

The following table provides bearing capacity factors based on the internal angle of friction. In between values may be interpolated:

ϕ (Degrees)	N_c	N_q	N_γ
0	5.7	1.0	0
5	7.3	1.6	0.5
10	9.6	2.7	1.2
15	12.9	4.4	2.5
20	17.7	7.4	5.0
25	25.1	12.7	9.7
30	37.2	22.5	19.7
34	52.6	36.5	35.0
35	57.8	41.4	42.4
40	95.7	81.5	100.4
45	172.3	173.3	297.5
48	258.3	287.9	780.1
50	347.5	415.1	1153.2

Shape Factors are based on the geometry of the footing where B is the width and L is the length as below:

B/L	S_c
1.0	1.25
0.5	1.12
0.2	1.05
Strip Footing	1.00
Circular	1.20

B/L	S_γ
1.0	0.85
0.5	0.90
0.2	0.95
Strip Footing	1.00
Circular	0.70

The ultimate bearing capacity then needs to be corrected for overburden to find the net bearing capacity:

$q_{net} = q_{ult} - \gamma D_f$

The allowable bearing capacity is then determined by dividing the net capacity by a predetermined factor of safety. A factor of safety of between 2 and 3 is common:

$q_a = q_{net}/FS$

Slope Stability

There are 3 types of slope failures:

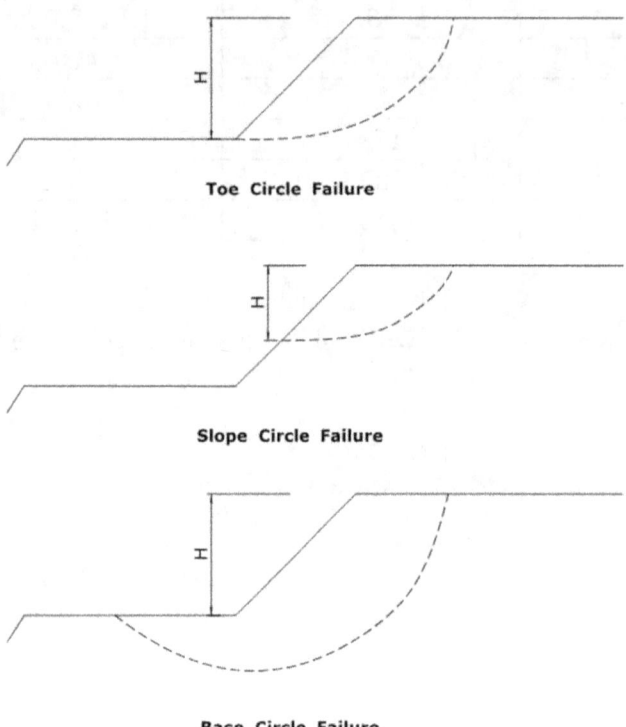

Toe Circle Failure

Slope Circle Failure

Base Circle Failure

$$F_{cohesive} = \frac{N_o c}{\gamma_{eff} H}$$

$F_{cohesive}$ = Safety factor for slope stability of cohesive soils. The minimum is often taken between 1.3-1.5
N_o = Stability number
c = cohesion (psi)
γ_{eff} = Effective soil density = $\gamma_{saturated} - \gamma_{water}$
H = Depth of cut

Structural

Dead and Live Loads

Dead Loads: Loads which are permanent in the final condition of the structure. Examples include self weight and additional permanent loads (such as pavement). Dead load factors are often lower than other types of loads. This is due to the higher level of reliability being able to predict the magnitude and character of these loads.

Live Loads: Loads which will or may change over time. In general, live loads represent pedestrian or vehicle loads. The load factor for live loads is often much higher due to the unpredictability.

In LRFD different types of loads are factored to represent a safety factor based on the reliability of our ability to accurately predict certain loading conditions. If only Dead and Live loads are present, the likely load combination is:

1.2D + 1.6L

Trusses

Trusses are structural members used to span long distances. Trusses are built up by members which are only in axial tension or axial compression. They can be analyzed by the method of joints as illustrated below. Consider the example truss with nodes labeled. To design, the axial force in each member must be determined. If we wanted to find the force in member BD, first like a typical beam, the reactions at A and B can be found by summing forces.

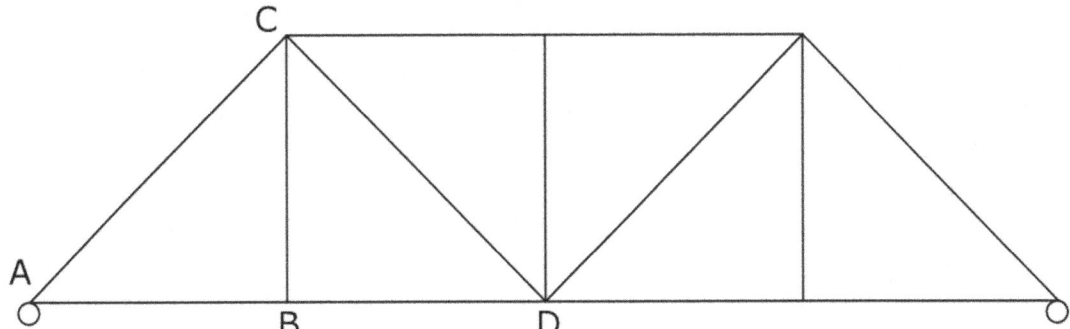

Then take a free body diagram of only the joint at A. This is shown below. In summing vertical reactions and since the reaction at A was found, the force in AC can be determined. Then there are only 2 horizontal forces and the force in AB can be found:

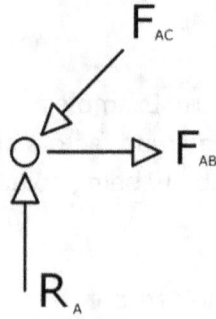

Then take a free body diagram of Joint B as shown below. Since there are only two horizontal forces, the axial force in member BD can then be determined:

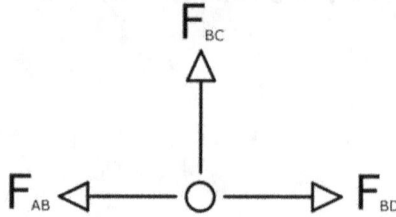

The remainder of the truss can be analyzed similarly.

Zero force members:

When determining how many 0-Force members a truss has, analyze each joint individually as a free body diagram and follow these guidelines:

1. In a joint with 2 members and no external forces or supports, both members are 0-force
2. In a joint with 2 members and external forces, If the force is parallel to one member and perpendicular to the other, then the member perpendicular to the force is a 0-force member.
3. In a joint with 3 members and no external forces, if 2 members are parallel then the other is a 0-force member

All other members are non-zero.

Bending Stress

Bending Stress

Mc/I

M = Applied Moment
c = Distance from the Centroid of the Cross Section to the Desired Location of Stress
I = Moment of Inertia of the Cross Section

Shear Stress

Shear stress at any point along a beam is the shear at that point over the area.

$\tau = V/A$

V = Shear at the point of interest
A = Cross sectional area

There is also horizontal shear stress due to bending

Horizontal shear stress $\tau = VQ/Ib$

V = Applied Shear Force (kips)
Q = First Moment of the Desired Area = ay.
a = Cross Sectional Area from Point of Desired Shear Stress to Extreme Fiber (in^2)
y = Distance from Centroid of Beam to Centroid of Area "a" (in)
I = Moment of Inertia of Beam (in^3)
b = Width of Member (in)

Axial Stress

Axial Stress:

P/A

P = Applied Force
A = Cross Sectional Area

Deflection

Deflection is the degree to which an element is displaced under load. Common equations for the maximum deflection of beams can be found in the Beam Chart

Beams

The chart below shows reactions, moments, and max deflections for common beam types:

Beam Type	Reaction	Maximum Moment	Maximum Deflection
Simply Supported w/ Single Point Load at Center	$\frac{P}{2}$	$\frac{PL}{4}$	$\frac{PL^3}{48EI}$
Simply Supported w/ Uniform Distributed Load	$\frac{wL}{2}$	$\frac{wL^2}{8}$	$\frac{5wL^4}{384EI}$
Cantilever w/ Single Point Load at Free End	P	PL	$\frac{PL^3}{3EI}$
Cantilever w/ Uniform Distributed Load	wL	$\frac{wL^2}{2}$	$\frac{wL^4}{8EI}$
One End Fixed, Supported at Other w/ Single Point Load at Center	$\frac{5P}{16}$, at Supported End $\frac{11P}{16}$, at Fixed End	$\frac{3PL}{16}$	$0.00932\frac{PL^3}{EI}$
One End Fixed, Supported at Other w/ Uniform Distributed Load	$\frac{3wL}{8}$, at Supported End $\frac{5wL}{8}$, at Fixed End	$\frac{wL^2}{8}$	$\frac{wL^4}{185EI}$
Beam Fixed at Both Ends, Single Point Load at Center	$\frac{P}{2}$	$\frac{PL}{8}$	$\frac{PL^3}{192EI}$
Beam Fixed at Both Ends, Uniform Distributed Load	$\frac{wL}{2}$	$\frac{wL^2}{12}$	$\frac{wL^4}{384EI}$

Shear and Moment Diagrams

Shear and moment diagrams are a graphical representation of the forces applied along the length of a beam. The following rules are used to develop shear diagrams:

-A concentrated force causes a jump in the shear diagram of equal magnitude
-A distributed load causes a line in the diagram with slope equal to the distributed load
-Forces up are positive and down is negative

This is depicted graphically below:

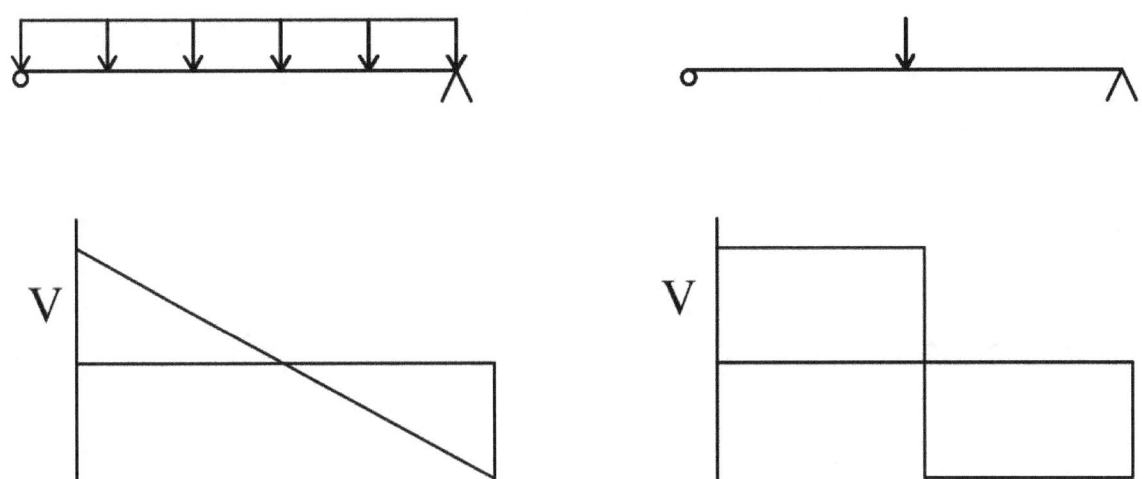

The following are rules for construction of a moment diagram:

-The moment at any point on the graph is equal to the area under the shear diagram up to this point
-An isolated moment causes a jump in the diagram of equal magnitude
-The shear at any point in the beam is equal to the slope of the same point on the moment diagram
-A distributed load will cause a parabolic moment diagram curve

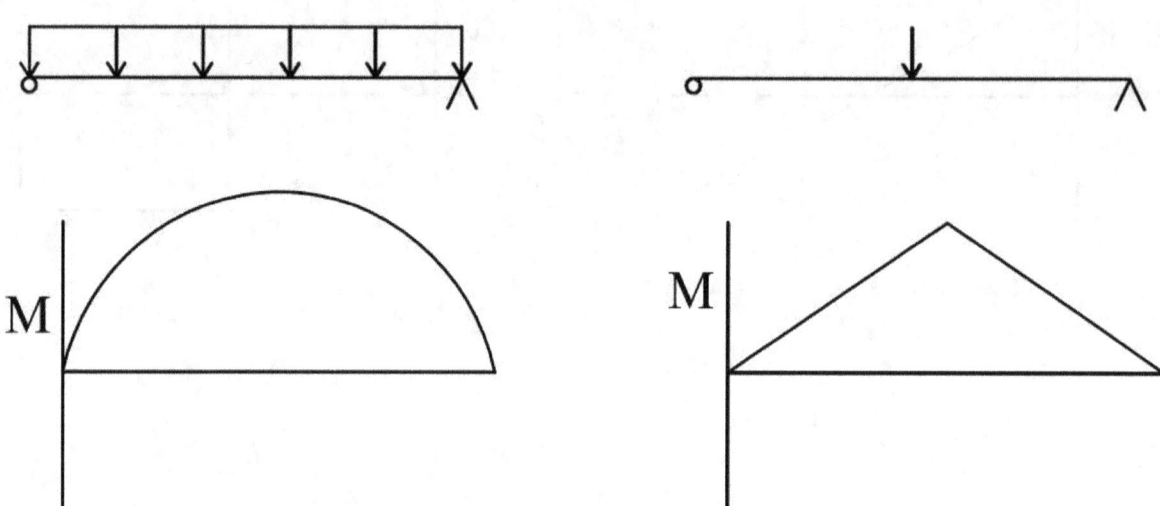

Columns

Columns are vertical members used to carry the load from spanning members to the foundation. For ideal columns where the load is concentric, the Euler formula is used to determine the theoretical maximum load:

Critical Load

$$P_{Cr} = \frac{\pi^2 EI}{(KL)^2}$$

E = Modulus of Elasticity (psi)
I = Moment of Inertia (in^4)
K = Effective Length Factor (See chart below)
L = Effective Length (in)

Critical Stress

$$F_{Cr} = \frac{\pi^2 E}{\left(\frac{KL}{r}\right)^2}$$

Effective Length Factor Chart

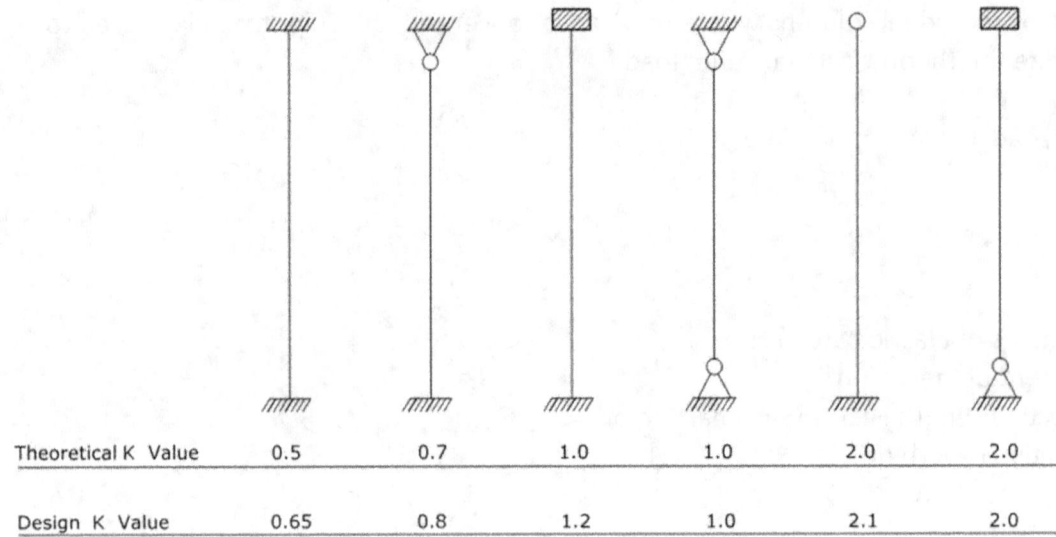

Theoretical K Value	0.5	0.7	1.0	1.0	2.0	2.0
Design K Value	0.65	0.8	1.2	1.0	2.1	2.0

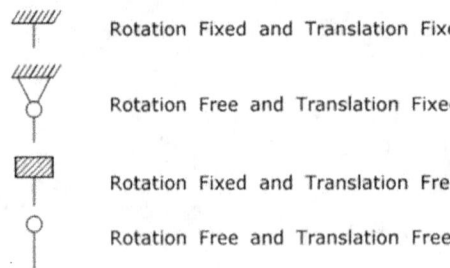

Slabs

One way slabs:

Slabs are structural elements whose length and widths are large compared to the thickness. Slabs are often used as floors or as foundation elements.

Flexure:

Slabs must be analyzed by simplified methods due to the indeterminacy of a full analysis. The most common of which is to analyze as a 1- foot wide strip and treat the span length as a beam. Transverse reinforcement is necessary to control temperature and shrinkage.

Shear:

Shear in slabs is also determined by taking one-foot wide sections to analyze as a beam. However often shear will not control.

Footings

This section will focus on shear and moment for wall footings and column spread footings.

One Way Shear:

The critical section for one-way shear is at a distance d from the support.

The force applied to the footing is assumed to spread uniformly therefore the bearing pressure q_u is the force divided by the area:

$$q_u = P/(Lb_w)$$

P = Factored Load
L = Length of Footing
b_w = Width of Footing

The shear force then is the shear resulting from the bearing pressure in the area from the critical section to the free end:

$$V_u = q_u b_w \left(\frac{L}{2} - \frac{a}{2} - d\right)$$

a = Width of Column
d = Depth to Flexural Reinforcing

Moment:

The critical section for moment is at the face of the column. The applied moment resulting from the bearing pressure is:

$$M_u = q_u \frac{\left(\frac{L}{2} - \frac{a}{2}\right)^2}{2} b_w$$

Retaining Walls

Retaining walls are built to facilitate an immediate change in elevation. Some uses are to support roadways or a need for a wide, level area to be formed from a sloping existing grade. Retaining walls are designed to resist lateral loads from active earth pressure (see geotechnical section for computation of these loads) and surcharge loads which is any additional load imposed on the soil above, which when close enough will cause an additional pressure on the load due to the distribution of this load through soil. The stem of retaining walls can be analyzed as a cantilever beam extending vertically from the footing. The footing is composed of the toe which is the portion on the side of the lower elevation of soil and the heel which is portion on the side of the higher elevation of soil. Retaining walls are analyzed on a per foot width:

$$Moment\ at\ base\ of\ stem = M_{stem} = R_{ah} y_a$$

R_{ah} = Horizontal Active Earth Pressure per ft Width
y_a = Eccentricity of Horizontal Active Earth Pressure

For shear, the critical section is a distance, d, from the base of the stem where d is the distance from the main flexural reinforcement (Heel side) to the extreme compression face (Toe side):

$$V_{stem} = R_{ah}$$

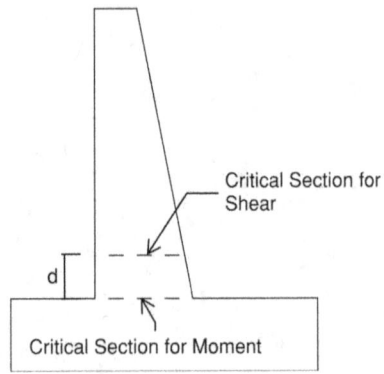

Hydraulics and Hydrology

All Hydraulics and Hydrology topics are covered in the afternoon section

Transportation and Geometrics

Horizontal curves

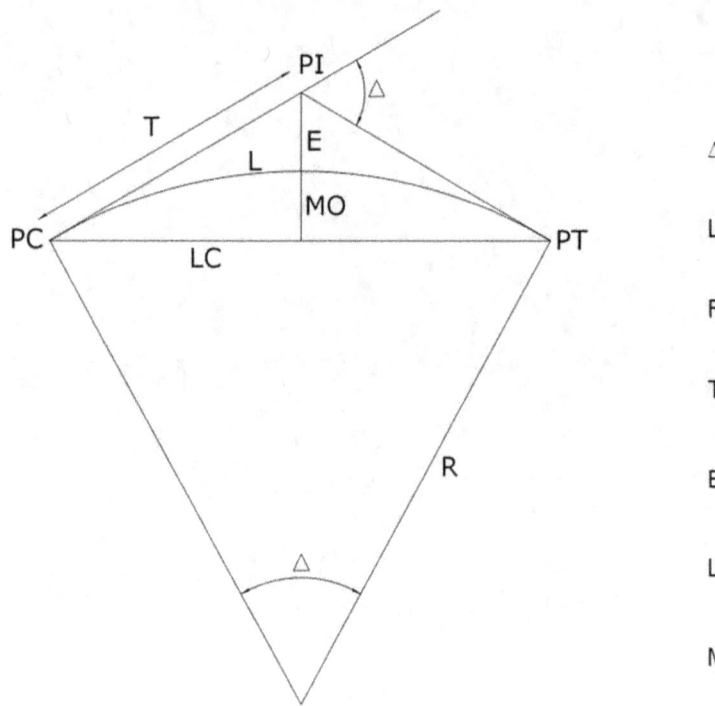

$$\Delta = \frac{180}{\pi} \times \frac{L}{R}$$

$$L = \frac{\Delta \times \pi \times R}{180}$$

$$R = \frac{180 \times L}{\Delta \times \pi}$$

$$T = R \times \left(\tan\frac{\Delta}{2}\right)$$

$$E = T \times \left(\tan\frac{\Delta}{4}\right)$$

$$LC = 2 \times R \times \left(\sin\frac{\Delta}{2}\right)$$

$$MO = R \times \left(1 - \cos\frac{\Delta}{2}\right)$$

Vertical Curves

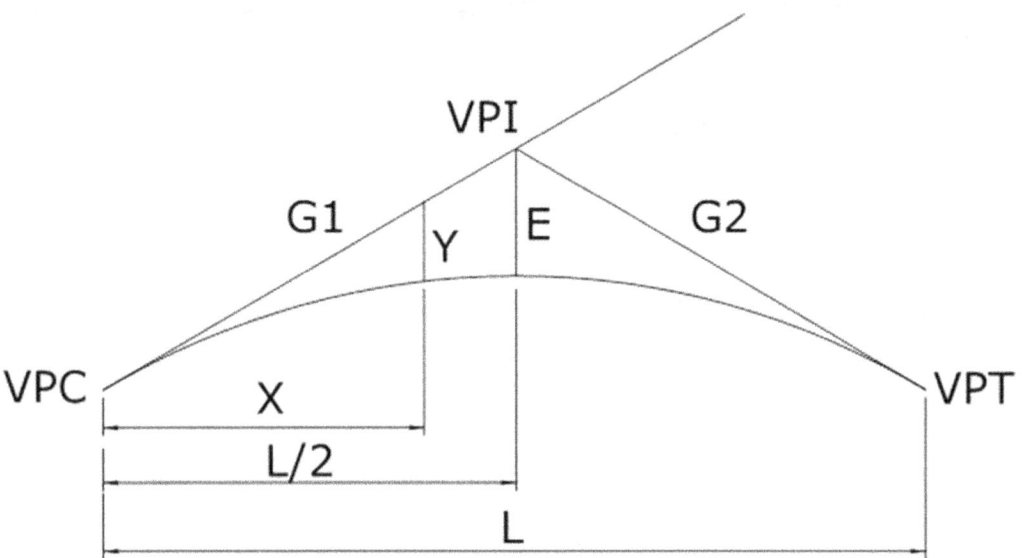

A = Gradient of the curve = $g_2 - g_1$
E = AL/800
Y = $Ax^2/200L$
Elevation at a distance x from VPC = $E_{VPC} + g_1x + ½(rx^2)$

Traffic Volume

The following are traffic volume factors. It is important to note for how many lanes and directions the values represent:

Average Daily Traffic, ADT: Average number of vehicles per day over a given time period.

Average Annual Daily Traffic, AADT: Average number of vehicles per day over a year. Typically it is calculated by dividing the total volume of vehicles in a year by 365 days.

Average Daily Truck Traffic, ADTT: Average number a trucks per day in a given time period.

Design Hour Volume, DHV: The hour of volume used in design.

K-factor, K: The ratio of the Design Hour Volume to the Average Annual Daily Traffic (DHV/AADT).

Directional Factor, D: Percentage of volume for the dominant direction of traffic during peak flow.

Directional Design Hour Volume, DDHV: The product of the Directional Factor and the Design Hour Volume.

Rate of Flow, v: Equivalent hourly rate at which vehicles pass a given point during a given time interval. The time frame is often taken as 15 min.

Design Capacity: Maximum volume a given roadway can handle.

Ideal Capacity, c: Ideal amount of volume for a given roadway. For freeways this is often taken as 2400 passenger cars per hour per lane (pcphpl).

Volume to capacity ratio: Volume over capacity, v/c

Peak Hour Factor, PHF: Ratio of the peak hour volume to the peak rate of flow in that hour:

$$PHF = \frac{V_{vph}}{4V_{15\ min,peak}}$$

Vehicle Dynamics

The distance it takes a driver to stop after recognizing an obstruction is the sum of two components. The first is before breaking and the second is after breaking.

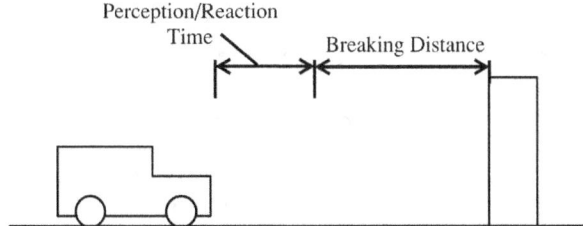

This is represented in the following equation:

$S_{stopping} = vt_p + s_b$

The first component assumes the velocity is constant during the perception reaction time which is the time it takes the driver to recognize the obstruction and begin the breaking process. This is calculated by:

vt_p where

v = Velocity (ft/s)
t_p = Perception Reaction Time (seconds)

The stopping distance during the breaking process is the following:

$s_b = v^2_{mph}/(30(f+G))$

v = Velocity (miles per hour)
f = Friction Factor with the Pavement
G = slope (positive for uphill and negative for downhill)

To convert the velocity from ft/s to mi/hour where applicable multiply by:

(3600 s/hr)/(5280 ft/mi)

Materials

Soil Classification

There are two types of common soil classification system. First is the AASHTO:

	Granular Materials (35% or less passing no. 200 sieve)							Silt-Clay materials (more than 35% passing no. 200 sieve)				
	A-1		A-3	A-2				A-4	A-5	A-6	A-7 / A-7-5 or A-7-6	A-8
	A-1-a	A-1-b		A-2-4	A-2-5	A-2-6	A-2-7					
Sieve analysis: % passing:												
no. 10	50 max											
no. 40	30 max	50 max	51 min									
no. 200	15 max	25 max	10 max	35 max	35 max	35 max	35 max	36 min	36 min	36 min	36 min	
Characteristics of fraction passing no 40:												
LL: liquid Limit				40 max	41 min	40 max	41 min	40 max	41 min	40 max	41 min	
PI: Plasticity Index	6 max		NP	10 max	10 max	11 min	11 min	10 max	10 max	11 min	11 min	
Usual types of significant constituents	Stone fragments gravel and sand		Fine Sand	Silty or clayey gravel and sand				Silty Soils		Clayey Soils		Peat, highly organic soils
General subgrade rating	Excellent to good							Fair to poor				Unsatisfactory

Second is the Unified Soil Classification System (USCS):

Major Division		Group Symbol	Laboratory classification criteria		Soil Description
			% finer than 200 sieve	Supplementary requirements	
Coarse-grained (over 50% by weight coarser than no. 200 sieve)	Gravelly soils (over half of coarse fraction larger than no. 4)	GW	0-5[a]	D_{60}/D_{10} greater than 4. D^2_{30} % / $(D_{60}D_{10})$ between 1 and 3	Well graded gravels, sandy gravels
		GP	0-5[a]	Not meeting above gradation requirement for GW	Gap-graded or uniform gravels, sandy gravels
		GM	12 or more[a]	PI less than 4 or below A-Line	Silty gravels, silty-sandy gravels
		GC	12 or more[a]	PI over 7 and above A-Line	Clayey gravels, clayey sandy gravels
	Sandy soils (over half of coarse fraction finer than no. 4)	SW	0-5[a]	D_{60}/D_{10} greater than 6. D^2_{30} % / $(D_{60}D_{10})$ between 1 and 3	Well-graded, gravelly sands
		SP	0-5[a]	Not meeting above gradation requirement for SW	Gap-graded or uniform sands, gravelly sands
		SM	12 or more[a]	PI less than 4 or below A-Line	Silty sand
		SC	12 or more[a]	PI over 7 and above A-Line	Clayey sands, clayey gravelly sands
Fine-grained (over 50% by weight finer than no. 200 sieve)	Low compressibility (liquid limit less than 50)	ML	Plasticity chart		Silts, very fine sands, silty or clayey fine sands, micaceous silts
		CL	Plasticity chart		Low plasticity clays, sandy or silty clays
		OL	Plasticity chart, organic odor or color		Organic silts and clays of high plasticity
	High Compressibility (liquid limit 50 or more)	MH	Plasticity chart		Micaceous silts, diatomaceous silts, volcanic ash
		CH	Plasticity chart		Highly plastic clays and snady clays
		OH	Plasticity chart, organic odor or color		Organic silts and clays of high plasticity
Soils with fibrous organic matter		Pt	Fibrous organic matter; will char, burn, or glow		Peat, sandy peats, and clayey peat

[a] For soils having 5-12% passing the no.200 sieve, use a dual symbol such as GW-GC

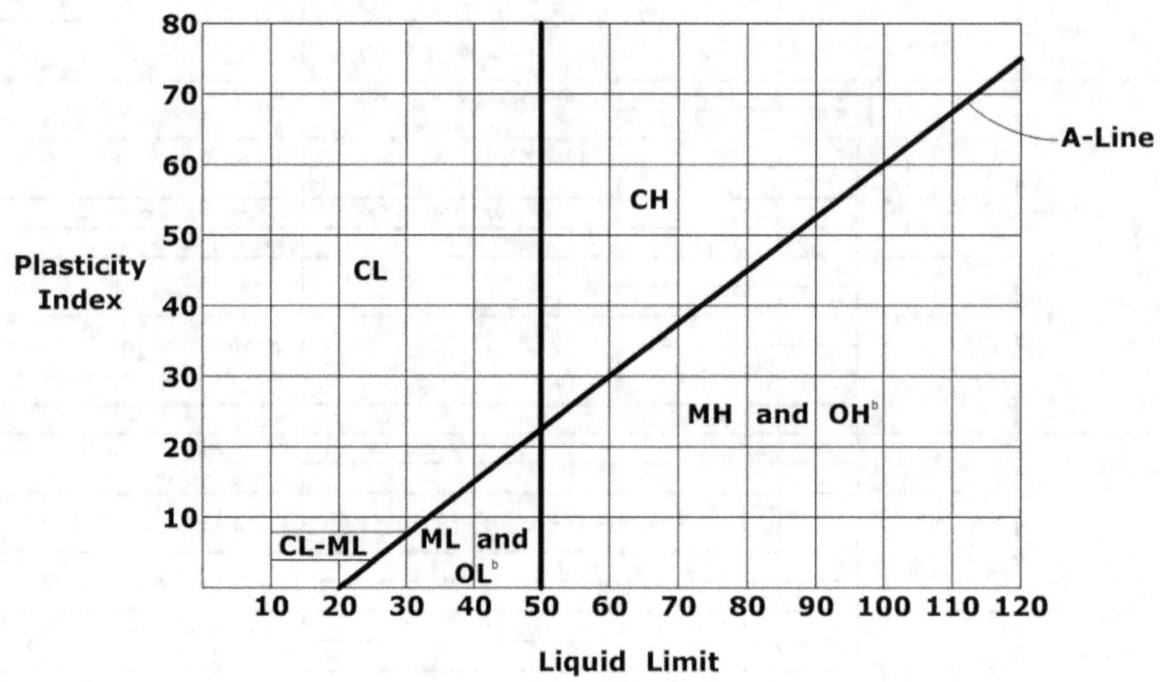

Soil Properties

The strength of soil is often determined by the standard penetration test. This measures the resistance to penetration using a standard split spoon sampler which is hit by a 140 lb hammer dropped from 30" high. The number of blows required to drive the sampler 12" after an initial penetration of 6" is referred to as the N-Value.

Phase Relationships

W_V = Weight of voids
W_w = Weight of water
W_s = Weight of solids
W_T = Total weight
V_V = Volume of voids
V_w = Volume of water
V_s = Volume of solids
V_T = Total Volume

Void ratio = $e = V_V/V_s$
Porosity = $n = V_V/V_T$
Degree of Saturation = $S = V_w/V_v \times 100\%$
Moisture content = $w = W_w/W_s$
Dry Unit weight = density = γ = Weight/Volume
SG = Specific Gravity = γ_s/γ_w
$\gamma_w = 62.4$ lb/ft^3 (constant)

A saturated sample indicates the volume of the voids = volume of water
A dry sample indicates the volume of voids includes no water

Concrete Properties

Concrete consists of cement, coarse aggregate, fine aggregate, and water. Additionally, concrete may contain admixtures to enhance a certain desirable aspect of the target product. Some properties of concrete include:

Concrete Strength, f'_c: The design compressive strength of the concrete. In general, this will range from 3000 to 6000 psi. However, strengths can be much higher such as 20,000 psi with proper mixing and additives.

Modulus of Elasticity (Normal weight concrete), $E_c = 57,000\sqrt{f'_c}$

Modulus of Rupture or the tensile strength, $f_r = 7.5\sqrt{f'_c}$ While the tensile strength of concrete is ignored in flexure, this is often used in cracking analysis.

Water to Cement ratio, w/c, is the amount of water to the amount of cement in a given mix. In general the w/c ratio is inversely proportional to the strength since the higher amount of cement, the stronger the mix.

Cement Types:

Type I – General use cement. When special properties are not desired Type I can be used.

Type II – Used in areas where sulfate attack is a concern. This is often in areas exposed to groundwater such as drainage structures. Type II will cure at a slower rate and therefore produce less heat than other types and gain strength at a slower rate.

Type III – High early strength concrete. As opposed to type II or IV, a large amount of heat is released quickly and therefore is not suitable for mass-type pours. Type III is used in concrete where rapid strength gain is desirable such as precast concrete.

Type IV – Low heat of hydration. Gains strength slowly and generates a low amount of heat. Often used for mass-pours such as mat foundations or large retaining walls.

Structural Steel Properties

Yield Strength, F_y: Stress at which the steel will yield and begin to cause permanent deformations.

Ultimate Strength, F_u: Stress at which the steel will fracture or fail in brittle behavior.

Modulus of Elasticity, E_s = The tendency of a material to deform when subjected to forces. Also is the ratio of stress over strain. Often in structural steel it is assumed to be taken as 29,000 ksi

Ductility: Measure of a materials ability to deform before failure. Ductility is the ratio of ultimate failure strain to yielding strain.

Toughness: The ability to withstand high stresses without fracturing.

Hardness: The ability of a material to resist surface deformation.

Material Test Methods and Specification Conformance

Concrete:

Strength tests: Most often strength is determined by loading cylinders often 6" in diameter to failure and recording the results.

Slump Test: Measure of the consistency and workability of a batch of concrete. A cone about 6" in diameter on the wide end and 12" tall is filled with concrete. The filled cone is placed on the ground and then removed to allow the concrete to naturally disperse. The remaining height and diameter of the concrete mix is measured and recorded.

Steel:

Tensile Test: Axially loading a steel member to recorded the strain in the member as the load increases. From this test the yield strength, ultimate strength, and stress strain curve can be determined. When the applied stress exceeds the yield strength, the member will undergo plastic deformation and the cross-sectional area will reduce until the member fractures. This is known as necking.

Fatigue Testing: Fatigue is damage caused by repeated cycles of loading. Even though the stress in fatigue is less than the yield strength of the member, the repetition over a long period of time can cause failure. A fatigue test measures the ability of a member to resist repeated cycles of stress at a given magnitude.

Scratch Hardness Test: Also known as Mohs Test. Compares the hardness of a material to that of minerals. Minerals of known and increasing hardness are used to scratch the sample and results are observed.

Charpy V-Notch Test: Measure of a member's toughness. A member is given a 45 degree notch and a pendulum is used to hit the opposite side of the member. This is performed at different heights and magnitudes until the member fails.

Compaction

Compaction is the reduction of voids in a mass of soil. The more compacted a mass of soil is, the more stable and strong it is to support a structure. Compaction is done by placing soil in layers called lifts and using equipment to mechanically apply weight and potentially vibration to the lifts. Some types of compaction equipment are Grid Rollers for rocky soil, sheep foot rollers for cohesive soils, or roller compactors with vibration capabilities for cohesion less soils.

When soil is compacted, the volume decreases. This is referred to as shrinkage. To calculate the compacted volume of a soil mass from its volume in its natural state use the following equation:

$$V_c = \left(\frac{100\% - \%shrinkage}{100}\right) V_b$$

V_c = Compacted Volume
%shrinkage = Percent Shrinkage
V_b = Volume of Soil in its Natural State

Conversely, when soil is removed from the ground there is an expansion of volume known as swelling:

$$V_l = \left(\frac{100\% + \%swell}{100}\right) V_c$$

V_l = Loose Volume
%swell = Percent Swell

Site Development

Excavation (Cut/Fill Estimates)

The most common method for determining the volume of excavation for cut and fill is the average end area method:

$$V = L(A_1+A_2)/2$$

L = Distance Between Area 1 and 2 (ft)
A_1 and A_2 = Respective Cross-Sectional Areas (ft^2)

Construction Site Layout

Construction sites are surveyed and markers are placed to indicate measurements and control points. These points are designated in the field by the use of stakes. These stakes can be called construction stakes, alignment stakes, offset stakes, grade stakes, or slope stakes depending on what they are meant to indicate. The accuracy of dimensions depends on the intent. Some accuracy requirements are shown below:

Type of Measurement	Level of Accuracy (ft)
Roadway Alignment, Intersections, and Paving	0.01
Buildings and Bridges	0.01
Culvert Lengths	0.1
Grade Staking	0.1
Culvert Stations	1.0
Telephone or Power Poles	1.0

Soil Erosion and Sediment Control

During construction activities involving excavation, there can be a significant amount of soil erosion leading to a dispersion of sediment. This needs to be controlled to prevent a negative impact to the surrounding areas. There are a number of options for sediment control:

Silt Fences: Fences consisting of a geotextile fabric and posts which allow the passing of runoff water but will catch the suspended sediment. They will be placed at the bottom of slopes and/or at the perimeter of the job sites at low points.

Hay Bales: Placed at the toe of slopes to help control runoff. Bales should be embedded in the ground and anchored securely with wooden posts.

Erosion Control Fabric: Geotextile fabric used for the control of erosion on steep slopes. Often these are used on piles of excavated material.

Temporary Seeding and Mulch: This involves seeding and mulching slopes to create growth that can control erosion due to the roots holding together soil. This is often used as a permanent measure for cut slopes.

Slope drain: A drain constructed to direct water to a specified area. The drain can be constructed with numerous materials such as plastic or metal pipes and concrete or asphalt. Drains must be properly anchored to resist forces from the flow of water. The outlet often is required to slow the flow of water by using energy dissipaters such as riprap.

Sediment Structure: An energy dissipating structure often made of rocks used to slow the flow of water and catch sediment.

Temporary Berm: A hill constructed of compacted soil to prevent runoff flowing in a specific direction. Berms are placed either at the top or bottom of slopes.

Impact of Construction on Adjacent Facilities

Construction can have a negative effect on surrounding properties and areas. These issues need to be anticipated and mitigated as possible. Some of the concerns include:

Construction Noise: OHSA sets maximum decibel limits on daily sound exposure. In the United States, this is typically 90 dBA for the eight hour noise level

Runoff and Sediment: Construction sites, especially those involving excavation, can change the dynamics of runoff and drainage. See the section on Soil Erosion and Sediment Control for more details.

View: Construction projects often change the landscape of the affected area. This may have an impact on the look and feel of an area. The needs of adjacent properties may need to be considered for these changes.

Rights of Way: Often, land which is not owned by the owner of the project is necessary for the final or temporary conditions. In these cases land needs to be acquired temporarily or permanently to complete the work. The owner of the project and of the land must come to an agreement to allow use of the property

Economic or social impact: Construction during and after may impact the access or desirability of a business or residential area. Consideration should be taken to limit the impacts to businesses or residents. For example, a bridge detour may cut off access to a restaurant which collects patrons mostly from tourists passing the effected route. The owner would then be compensated for the loss of business.

Safety

Safety is extremely important for construction sites. The *OSHA CFR 29 Part 1910 and Part 1926: Occupational Safety and Health Administration* provides requirements for all types of construction situations and is recommended to use for the exam. Some of the highlights which you should further familiarize yourself with include:

- Excavation Safety: Except for excavations in rock, anything deeper than 5 ft must be stabilized to prevent cave-in. This may be achieved by providing appropriate earth retention systems or sloping at appropriate rates. This is determined by the depth of excavation, soil type, and other requirements.
- Fall protection: Drop-offs must be protected from fall based on the height of the drop. Some examples of protection include temporary fences, nets, or lifelines.
- Roadside Safety: Construction sites adjacent to traffic must be sufficiently protected from impact. At higher speeds concrete barriers may be needed also known as temporary precast concrete barrier curbs (TPCBC). At lower speeds it may be acceptable to provide barrels or cones to delineate the work area.
- Power line Hazards: For power lines which are electrified, all construction activities must be a minimum distance from the lines. This is based on the voltage of the lines. Typically the safe operational distance is 10 ft. for lines less than 50 kV and typically 35 ft. for lines greater than 50 kV.
- Confined Spaces: Anyone required to enter confined spaces must be appropriately trained and equipped. Oxygen must be monitored and kept at an acceptable level.
- Personal Protective Equipment (PPE): Equipment required by any personnel present on a job site. The main aspects are acceptable head protection and steel toed shoes.

I. Analysis and Design

A. Mass balance

Mass balance refers to the conservation of mass as it enters and exits a system. This concept is essential to the Water Resources and Environmental exam as it is a common method in solving for unknowns in many types of problems. This principle simply put is that mass is always conserved in fluid systems regardless of the properties of that system. Therefore, what enters the system must also exit the system. The equation can be represented as mass:

$M_1 = M_2$

M_1 = Mass entering the system
M_2 = Mass exiting the system

The equation can also be represented by equating flow rates. In this case the equation is commonly known as the continuity equation:

$Q_1 = Q_2$

Q_1 = Flow rate into system
Q_2 = Flow rate out of a system

Given that the flow rate is equal to the area multiplied by the velocity, the equation can be used to equate these variables as well:

$A_1 V_1 = A_2 V_2$

B. Hydraulic loading

Hydraulic loading refers to the flows in MGD (Million Gallons per Day) or cu. Ft./day to a treatment plant or treatment process. The equation is as follows:

$$H_{LR} = \frac{Q}{A}$$

H_{LR} = Hydraulic loading rate
Q = Flow rate
A = Surface area of the wet basin

Detention time is the amount of time it takes a given volume of wastewater to pass through the clarifier:

$$t_d = \frac{V}{Q}$$

V = Volume of clarifier

C. <u>Solids loading (e.g., sediment loading, sludge)</u>

Solids loading similarly to hydraulic loading is the amount of suspended solids in a substance as it flows to the treatment facility. Solids loading is expressed as the following:

$$\text{Solids Loading Rate} = \frac{\text{Suspended Solids}\left(\frac{lb}{day}\right)}{\text{Surface Area (ft}^2)}$$

D. <u>Hydraulic flow measurement</u>

Many flow devices are available to measure either the flow rate or velocity of a given system. These methods are often used in conjunction with the laws of energy and mass conservation to be able to analyze a system. Here are a few of the more prevalent ones for the purposes of the PE exam:

1. Pitot Static Tubes

$$v = \sqrt{\frac{2gh(\rho_m - \rho)}{\rho}}$$

g = Force effect due to gravity (32.2 ft/s²)
h = Difference in elevations of the fluid columns (ft)
ρ = Density of water (62.4 lb/ft³)
ρ_m = Density of manometer fluid (lb/ft³)
v = Velocity (ft/s)

2. Orifice or Venturi Meter

$$\text{Flow rate through an orifice meter} = Q = C_f A_o \sqrt{\frac{2g(\rho_m - \rho)h}{\rho}} = C_f A_o \sqrt{\frac{2g_c(p_1 - p_2)}{\rho}}$$

$$C_f = C_d F_{va}$$

$$F_{va} = \frac{1}{\sqrt{1 - \left(\frac{C_c A_o}{A_1}\right)^2}}$$

C_f = Flow coefficient
A_o = Orifice area
A_1 = Pipe area
p = Pressure
ρ = Density
g = Force due to gravity
C_d = Discharge coefficient
C_c = Coefficient of contraction
F_{va} = Velocity of approach factor

II. Hydraulics–Closed Conduit

A. <u>Energy and/or continuity equation (e.g., Bernoulli, momentum equation)</u>

The Bernoulli equation for the conservation of energy states that the total energy is equal to the sum of the pressure + kinetic energy + potential energy of a system and is conserved at any point in the system. Therefore:

$E_t = E_{pr} + E_v + E_p = p + v^2/2g + z$

E_{pr} = Pressure = p
E_v = Kinetic Energy = $v^2/2g$
v = Velocity (ft/s)
g = Acceleration Due to Gravity (32.2 ft/s²)
E_p = Potential Energy = z = Height above point of interest to surface of water (ft)

B. <u>Pressure conduit (e.g., single pipe, force mains, Hazen-Williams, Darcy-Weisbach, major and minor losses)</u>

Pressure conduits refer to closed cross sections that are not open to the atmosphere such as pipes:

The Darcy Equation is used for fully turbulent flow to find the head loss due to friction. The equation is:

$h_f = (fLv^2)/(2Dg)$

h_f = Head Loss due to friction (ft)
f = Darcy friction factor
L = Length of pipe (ft)
v = Velocity of flow (ft/sec)
D = Diameter of pipe (ft)
g = Acceleration due to gravity, (Use 32.2 ft/sec^2)

The Hazen-Williams equation is also used to determine head loss due to friction. Be aware of units as this equation may be presented in different forms. The most common is the following:

$h_f = 10.44 L V^{1.85}/C^{1.85} d^{4.87}$

h_f = Head Loss due to F\friction (ft)
L = Length (ft)
V = Velocity (gallons per minute)
C = Roughness coefficient
d = Diameter (in)

In addition to these losses, there is also what is called minor losses of energy due to friction

Minor Losses – Friction losses due to geometric changes such as fittings in the line, changes in the dimensions of the pipe, or changes in direction:

- Minor losses can be calculated as per the Method of Loss coefficients.
- Each change in the flow of a pipe is assigned a loss coefficient, K
- Loss coefficients for fittings are most often determined and provided by the manufacturer

Loss coefficients for sudden changes in area can be determined:

For Sudden Expansions:

$$K = \left(1 - \left(\frac{D_1}{D_2}\right)^2\right)^2$$

For Sudden Contractions:

$$K = 0.5\left(1 - \left(\frac{D_1}{D_2}\right)^2\right)$$

$D_1 = Smaller\ diamter\ pipe$
$D_2 = Larger\ diamter\ pipe$

Loss coefficients are then multiplied by the kinetic energy to determine the loss:

$$h_f = K\frac{v^2}{2g}$$

C. Pump application and analysis, including wet wells, lift stations, and cavitation

A pump is a machine which adds energy to the flow of water or other fluids. A pump is often used to oppose the effects of gravity to transport a fluid to a position up grade.

The head added by a pump can be determined from the following equation as a function of the total energy:

$$h_a = \frac{E_a g_c}{g}$$

To determine the power needed by a pump from the energy, the following equation can be applied:

$$h_a = \frac{\left(550 \frac{ft-lb}{sec-hp}\right) P_{hp,input} \eta_{pump}}{Q\gamma}$$

h_a = Head added (ft)
Q = Flow rate (ft³/s)
γ = Density of fluid (lb/ft³)
P = Power (Horsepower)

The water horse power can also be determined from the power:

$$WHP = P_{hp,input} \eta_{pump}$$

η_{pump} = Pump Efficiency

D. Pipe network analysis (e.g., series, parallel, and loop networks)

A system of pipes can be arranged in different configurations to be able to appropriately transport water. There are a few types of common arrangements that can be used. Each has certain principles to follow when determining the flow through the system. It is important to remember the conservation of mass or flow principle when analyzing these systems:

Series Pipe System: Pipes of different areas connected along the same line.

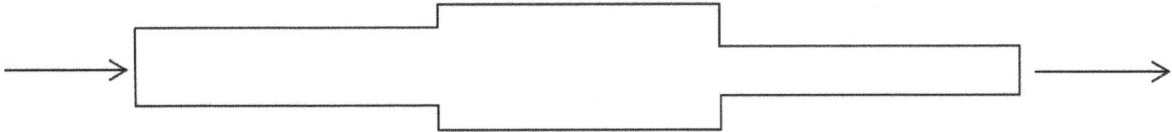

In a series pipe, the total friction loss is the sum of the loss in all the individual pipes. Therefore, in a pipe such as the one shown above the total head loss can be determined as follows:

$$h_{f,t} = h_{f,1} + h_{f,2} + h_{f,3}$$

Parallel Pipe Systems: As the name suggests, this is a pipe system with flow separating into parallel pipes.

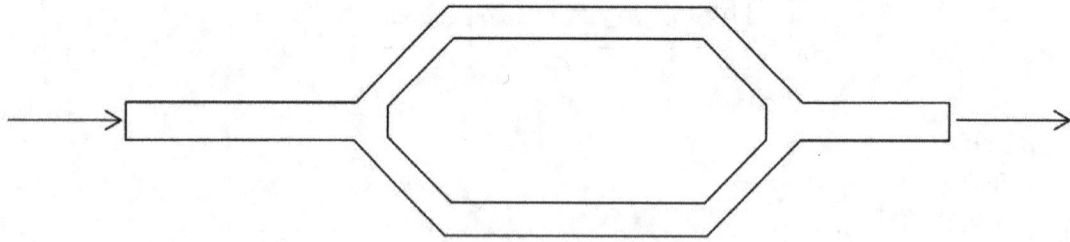

There are three concepts which are important to keep in mind during the analysis of parallel pipes:

1. The head loss in parallel pipes is equal
2. The head loss between the inlet and outlet is equal to that of each pipe individually
3. The flow rate at the outlet is equal to the sum of the flow rates from the parallel pipes

Pipe Networks: These are more complicated systems of pipes which have flow breaking off in multiple directions.

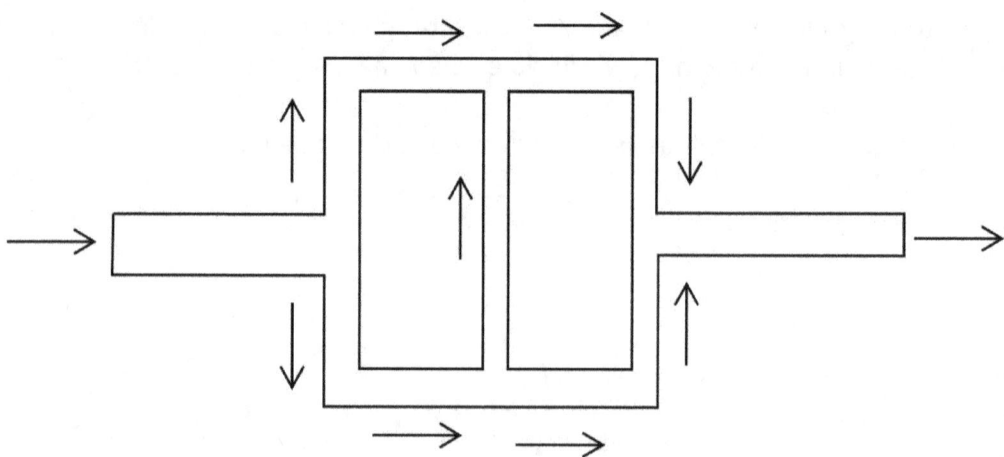

Often pipe networks are very complicated and left to iterative analysis on computers. It is important to note the two concepts which govern the analysis however:

1. The flow entering the system is equal to the flow leaving the system (conservation of flow)
2. The sum of head losses in any closed loop is equal to zero

III. Hydraulics–Open Channel

A. Open-channel flow

For open channel flow use the Chezy-Manning equation:

$Q = (1.49/n)AR^{2/3}S^{1/2}$

Q = Flow Rate (cfs)
n = Roughness Coefficient
A = Area of Water (ft²)
R = Hydraulic Radius (ft)
S = Slope (decimal form)

The hydraulic radius is the area of water divided by the wetted perimeter which is the perimeter of the sides of the channel which are in contact with water.

B. Hydraulic energy dissipation

A weir is a low dam used to control the flow of water. Weirs have shaped outlets notched into the top of the dam to allow water to flow out. The most common shapes are triangular and trapezoidal:

Triangular Weir

$$Q = C_2 \left(\frac{8}{15} tan\left(\frac{\theta}{2}\right)\right)\sqrt{2g}H^{5/2}$$

$$Q = 2.5H^{5/2} \ (For\ 90^o\ weir)$$

H = Height of water (ft)
θ = Weir angle

Trapezoidal Weir

$$Q = \frac{2}{3}C_d b\sqrt{2g}H^{3/2}$$

Often, the weir can be approximated by taking C_d, the discharge coefficient = 0.63 and the equation is simplified as:

$$Q = 3.367bH^{3/2}$$

b = Width of base (ft)

Broad Crested Weirs (Spillways)

Spillways are used to control the flow of excess water from a dam structure. Essentially they are large weirs and therefore can be called broad crested weirs. The calculation of discharge for spillways is taken as:

$$Q = C_s b \left(H + \frac{v^2}{2g}\right)^{3/2}$$

However, often in a dam situation the approach velocity can be taken as zero since it is so small and the equation becomes:

$$Q = C_s b (H)^{3/2}$$

C_s = Spillway coefficient

C. Stormwater collection and drainage (e.g., culvert, stormwater inlets, gutter flow, street flow, storm sewer pipes)

There are many components used in the collection of stormwater. Some examples include:

Culverts: A pipe carrying water under or through a feature. Culverts often carry brooks or creeks under roadways. Culverts must be designed for large intensity storm events.

Stormwater Inlets: Roadside storm drains which collect water from gutter flow or roadside swales.

Gutter/Street flow: Flow which travels along the length of the street. Gutter flow can be approximated often by an adaptation of the Manning Equation:

$$Q = K(^z/_n)s^{1/2}y^{8/3}$$

K = Gutter flow constant = 0.56 ft^3/(s*ft)
z = Inverse of the cross slope of the gutter (Decimal)
n = Roughness coefficient
s = Slope of the gutter (Decimal)
y = water depth at the curb (ft)

Storm Sewer Pipes: Pipes installed under the road which carry the water from inlets to a suitable outlet.

E. Sub- and supercritical flow

Sub and super critical flows are relative to the critical flow depth. This depth is defined as that which minimizes the energy of the flow of water for a given channel cross section and slope. The critical flow depth is important because since the energy is minimized the flow rate is maximized.

So when the flow depth is greater than the critical depth, the flow is subcritical and the velocity is less than the critical velocity. When the depth is less than the critical depth, the flow is supercritical and the velocity is faster than the critical.

For a rectangular channel, the following equations can be used to determine the critical depth and critical velocity:

$$d_c^3 = \frac{Q^2}{gw^2}$$

$$v_c = \sqrt{gd_c}$$

d_c = Critical depth of flow (ft)
Q = Flow rate (ft^3/s)
w = Width (ft)

The Froude number is used to qualify a flow channel and can be used to determine if it is sub or supercritical. The number is dimensionless:

$$Fr = \frac{v}{\sqrt{gL}}$$

v = Velocity (ft/s)
L = Characteristic length and is determined based on the channel geometry.

For a rectangular section L = d
For a circular section flowing half full L = πD/8
For trapezoidal and semi-circle sections L = the area of flow/top width of channel

If the Froude number is **less than 1**, the flow is subcritical
If the Froude number is **greater than 1,** the flow is supercritical

A hydraulic jump is a rise in water elevation due to a supercritical flow abruptly meeting a subcritical flow. The following equations can determine the heights and velocities in a hydraulic jump for rectangular sections:

$$d_1 = -\frac{1}{2}d_2 + \sqrt{\frac{2v_2^2 d_2}{g} + \frac{d_2^2}{4}}$$

$$\frac{d_2}{d_1} = \frac{1}{2}\left(\sqrt{1 + 8(Fr_1)^2} - 1\right)$$

$$v_1^2 = \left(\frac{gd_2}{2d_1}\right)(d_1 + d_2)$$

d_1 = depth of supercritical flow (ft)
d_2 = depth of subcritical flow (ft)
Fr = Froude Number for associated depth
v = Velocity at associated depth (ft/s)
g = Force effect due to gravity (32.2 ft/s²)

IV. Hydrology

A. Storm characteristics (e.g., storm frequency, rainfall measurement, and distribution)

A design storm must be specified when performing any calculations. The design storm is defined by its recurrence interval which is the given amount of time it is likely to see a storm of a certain intensity. Design storms are often 10, 20, 50, or 100-year storms meaning a storm of a certain intensity would only occur once within the given duration.

B. Runoff analysis (e.g., Rational and SCS/NRCS methods)

The rational method can be used to determine the flow rate from runoff of a drainage area. The equation is:

Q = ACi

Q = Flow Rate (cfs)
A = Drainage Area (Acres)
C = Runoff Coefficient
i = Rainfall Intensity (in/hr)

NRCS/SCS Runoff Method

This is an alternative method for determining runoff:

$$S = \frac{1000}{CN} - 10$$

S = Storage Capacity of Soil (in.)
CN = NRCS Curve Number

$$Q = \frac{(P_g - 0.2S)^2}{P_g + 0.8S}$$

Q = Runoff (in.)
P_g = Gross Rain Fall (in.)

C. Hydrograph development and applications, including synthetic hydrographs

Hyetographs – Graphical representation of rainfall distribution over time

Hydrograph – Graphical representation of rate of flow vs time past a given point often in a river, channel, or conduit. The area under the hydrograph curve is the volume for a given time period

Parts of a Hydrograph are shown graphically:

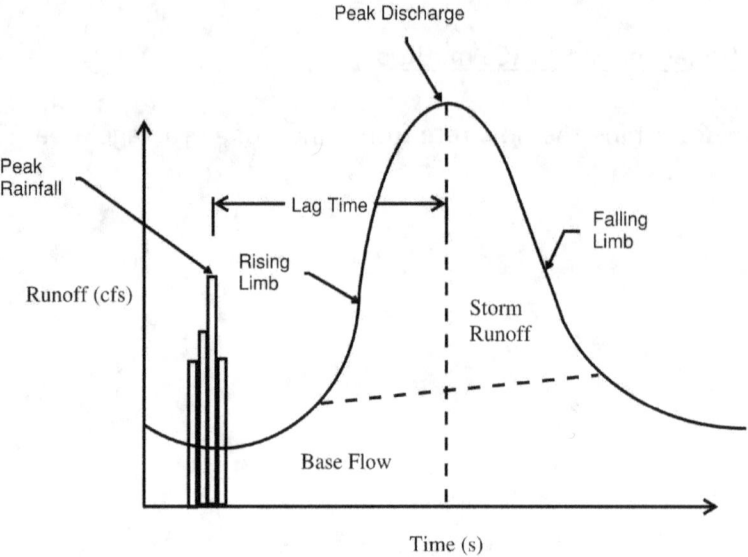

Unit Hydrographs can be determined by dividing the points on the typical hydrograph by the average excess precipitation.

Synthetic Hydrographs are created if there is insufficient data for a watershed. This method uses the NRCS curve number and is a function of the storage capacity.

$$S = \frac{1000}{CN} - 10$$

To develop the synthetic hydrograph, you must calculate the time to peak flow:

$$t_p = 0.5 t_R + t_1 \text{ where } t_1 = \frac{L_o^{0.8}(S+1)^{0.7}}{1900\sqrt{S_{percentage}}}$$

t_R = Storm duration (time)
L_o = Length overland (ft)
$S_{Percentage}$ = Slope of land

The equation for peak discharge from a synthetic hydrograph then is:

$$Q_p = \frac{0.756 A_{d,Acres}}{t_p}$$

D. Rainfall intensity, duration, and frequency

Storm characteristics include duration, total volume, and intensity

Duration: The length of time of a storm. Often measured in days and hours.

Total Volume: The entire amount of precipitation throughout the duration of the storm in a defined area.

Storm Intensity: Total volume of the storm divided by the duration of the storm event. Intensities can be averaged over the entire storm or at shorter intervals to provide instantaneous high intensity portions of the storm. Hyetographs are bar graphs used to measure instantaneous rainfall intensities over time and are covered below.

E. Time of concentration

Time of Concentration, t_c: The time of travel for water to move from the hydraulically most remote point in a watershed to the outlet. The time of concentration is the sum of three components:

$$t_c = t_{sheet} + t_{shallow} + t_{chann}$$

First, water moves as sheet flow:

$$t_{sheet} = \frac{0.007(nL_o)^{0.8}}{\sqrt{P_2} S_{deciaml}^{0.4}}$$

n = Manning's roughness coefficient
L_o = Length overland (ft)
P_2 = 2-yr, 24-hour rainfall (in.)
$S_{decimal}$ = Slope of the hydraulic grade (Decimal Form)

Flow then becomes shallow concentrated flow and typically enters a swale or ditch:

$$t_{shallow} = \frac{L_{shall}}{v_{shallow}}$$

$$v_{shallow} = 16.1345\sqrt{S_{decimal}} \text{ for unpaved flow}$$

$$v_{shall} = 20.3282\sqrt{S_{decimal}} \text{ for paved flow}$$

$L_{shallow}$ = Length of concentrated flow (ft)

$V_{shallow}$ = Velocity of concentrated flow (ft/s)

Flow finally will then enter a storm drain or channel:

$$t_{chann} = \frac{L_{chann}}{v_{chann}}$$

$L_{channel}$ = Length of channel (ft)

$V_{channel}$ = Velocity of Channel (ft, often determined by Manning or Hazen-Williams Equation)

F. Rainfall and stream gauging stations

Stream gauging is the measurement of a stream channel to determine the discharge by obtaining the depth and velocity of the channel over time. The channel can be approximated by areas created by connected the dots of the measured depths. The discharge can be calculated by the following:

$$Q = w\left(\frac{y_1 + y_2}{2}\right)\left(\frac{v_1 + v_2}{2}\right)$$

w = Width of cross section (ft)
y = Height of cross section (ft)
v = Velocity at indicated cross section (ft/s)

G. Depletions (e.g., evaporation, detention, percolation, and diversions)

The change in storage for a body of water can be approximated from the following equation:

$$\Delta S = P + R + GI - GO - E - T - O$$

S = Storage
P = Precipitation
R = Runoff
GI = Groundwater inflow
GO = Groundwater outflow
E = Evaporation
T = Transpiration
O = Surface water release

H. Stormwater management (e.g., detention ponds, retention ponds, infiltration)

Detention and retention ponds are often used to collect water for flood control and stormwater runoff treatment.

Detention Ponds: Also known as dry ponds. These are ponds which are often kept dry except during flood events. The pond will fill up during increased precipitation to control the flow intensity. This is common in dry, arid, or urban areas to prevent excessive flooding. The ponds typically will be designed to hold water for about 24 hours. Detention ponds also control the amount of sediment since it is captured in the pond and then typically becomes accessible after the pond has dried.

Retention Ponds: Also called wet ponds since they contain a volume of water at all times. The elevation of the water will vary depending on precipitation but will always maintain a permanent amount of water based on low flow conditions. This allows sediment control since the deposits will settle to the bottom and allow for collection.

Infiltration is the rate of which water seeps into the ground. The Horton equation can be used to approximate this rate. This assumes that the water supply is infinite and the ground is saturated:

$$f_p = f_c + (f_o - f_c)e^{-kt}$$

f_p = Infiltration rate at time t (in/hr)
f_c = Ultimate infiltration rate (in/hr)
f_o = Initial infiltration rate (in/hr)
k = Decay constant (time^{-1})
t = Time (hr)

V. Groundwater and Wells

A. Aquifers

Aquifers are bodies of saturated rock which contain and allow for the flow of water. Aquifers then by definition must be of a material that is porous and permeable. Some examples of material which are common for aquifers are sandstone, fractured limestone, and gravel but many others exist. Aquifers can be divided into 2 zones separated by the portion above and below the water table. The portion below is also saturated whereas that above may be fully, partially, or not saturated.

Aquifers are often defined by certain characteristics which can be calculated using soil samples. These are the same as discussed in the soil properties section of the morning session but we define some of the most common characteristics here:

$$Moisture\ Content\ w = \frac{m_w}{m_s} = \frac{m_t - m_s}{m_s}$$

$$Porosity\ n = \frac{V_v}{V_t} = \frac{V_t - V_s}{V_t}$$

$$Void\ Ratio\ e = \frac{V_v}{V_s} = \frac{V_t - V_s}{V_s}$$

The Hydraulic Gradient is also an important characteristic in defining an aquifer. This is defined as the change in hydraulic head over a particular distance:

$$i = \frac{\Delta H}{L}$$

i = Hydraulic gradient
H = Height (ft)
L = Length (ft)

B. Groundwater flow

The flow of Groundwater is defined by Darcy's Law. The flow rate of water through soil depending on its permeability can be measured by Darcy's Law:

$Q = KiA_{gross}$

Q = Flow rate (cfs)
K = Coefficient of permeability (ft/sec)
i = Hydraulic gradient

The coefficient of permeability can be determined from the following equation:

$$K = \frac{k\gamma}{\mu}$$

k = Intrinsic permeability (ft²)
μ = Absolute viscosity (lb-s/ft²)
γ = Specific weight (lb/ft³)

D. Well analysis–steady state

Discharge for a well in a confined aquifer is determined from the Thiem equation. Note this is only for steady flow:

$$Q = \frac{2\pi KY(y_1 - y_2)}{\ln\frac{r_1}{r_2}}$$

For an unconfined aquifer, the discharge is determined from the Dupruit Equation:

$$Q = \frac{2\pi K(y_1^2 - y_2^2)}{\ln\frac{r_1}{r_2}}$$

K = Hydraulic conductivity (ft³/(day-ft²))
Y = Original aquifer thickness (ft)
y_1, y_2 = Aquifer depths at distances r_1 and r_2 (ft)
r = Radial distance from well (ft)

VI. Wastewater Collection and Treatment

A. Wastewater collection systems (e.g., lift stations, sewer networks, infiltration, inflow, smoke testing, maintenance, and odor control)

Wastewater is collected by a network of pipes known as sanitary sewer. Some residents may not be connected to the sewer line and may resort to septic tanks for their wastewater disposal. Wastewater in sewer networks is transported to the wastewater treatment plants to be treated. Lift stations are pump stations which can be used to facilitate the transportation of the waste water to the treatment plant.

Smoke testing is a method of determining if there are leaks in a wastewater system. Smoke is pumped into a pipe and will seep through any cracks which can then be identified.

Infiltration is water which enters the system due to imperfections in the system such as cracks in the line or improperly constructed portions.

Inflow is water that enters the system from unanticipated or illegal means.

B. Wastewater treatment processes

Wastewater treatment processes are the procedures for treating wastewater so that it may be used again. This process will remove sediments, sludge, taste, odors, and any other undesirable characteristics of the water. The process can be divided into preliminary, primary, and secondary treatment which will be discussed further below.

C. Wastewater flow rates

The quantity of wastewater from a municipal needs to be determined to properly design the treatment system. This is based on anticipated discharge from residential, commercial and other buildings. In addition, the system must account for infiltration and inflow as defined previously.

Flow rate can be approximated as the average flow or the peak flow. The peak flow is the highest daily flow rate. The average and peak flow are related by the peaking factor:

$$\frac{Q_{Peak}}{Q_{avg}} = PF$$

The peak factor can also be approximated by the population using the Harmon equation:

$$PF = \frac{18 + \sqrt{P}}{4 + \sqrt{P}}$$

P is the population in thousands of people

D. Preliminary treatment

Preliminary treatment is the first step in the wastewater treatment process. This portion of the process is mostly the mechanical removal of debris and other large objects which may be caught in the flow. Heavy chemicals and large amounts of oil are also removed during this process. In general, anything that can be identified with the naked eye and easily screened will be removed during the preliminary treatment process. This process is often performed with large mechanical screens or filters. These large obstructions must also be removed so that they do not damage or impede the subsequent processes.

E. Primary treatment

Primary treatment is the second level in wastewater treatment. In this portion the wastewater is allowed to settle to remove any remaining oils and any solids which are able to separate. Typically, about half of the solids will be removed during this portion of the process. It is also expected that this level of the process will remove 25%-35% of the Biochemical Oxygen Demand (BOD) in the wastewater.

F. Secondary treatment (e.g., physical, chemical, and biological processes)

The most intensive of the levels of wastewater treatment is the secondary treatment. This may involve biological treatment in tickling filters and sludge treatment. The most amount of BOD will be removed in this stage.

G. Nitrification/denitrification

Nitrification is the use of oxygen by autotrophic bacteria. In this process the bacteria oxidizes ammonia to nitrites and nitrates. This process is important to understand as it relates to determining the Biochemical Oxygen Demand (BOD) of a particular sample and more specifically, the Ultimate Biochemical Oxygen Demand (BOD_u). To test for BOD, samples are diluted and dissolved oxygen is measured initially and typically after a 5-day period. The following equation is used to determine the Biochemical Oxygen Demand after that 5-day period (BOD_5):

$$BOD_5 = \frac{DO_i - DO_f}{\frac{V_{Sample}}{V_{Sample} + V_{Dilution}}}$$

DO_i = Initial Dissolved Oxygen content
DO_f = Final Dissolved Oxygen content
V = Volume

The process of nitrification causes a deviation from the trajectory of the carbonaceous process of oxygen demand as it relates to time. This must be accounted for when determining the Ultimate BOD. The BOD at any time t is called the BOD exertion and is related to the ultimate by the following equation:

$$BOD_t = BOD_u(1 - 10^{-K_d t})$$

It is important to note that at initially in a sample there is only a small amount of autotrophic bacteria present and the process of nitrification is delayed from having a significant effect on the BOD process. For reference the chemical equation for nitrification is:

$$NH_4^+ + 2O_2 \rightarrow NO_3^- + H_2O + H^+$$

Denitrification is the removal or loss of nitrogen by the means of bacteria. The chemical equation is:

$$NH_4^+ + OH^- \rightarrow NH_3 + H_2O$$

H. Phosphorus removal

Phosphorous removal can be separated into two different types. A small percentage is insoluble and can be removed during primary settling. The remaining amount is soluble and must be chemically converted to an insoluble material for removal.

Often aluminum sulfate, ferric sulfide, and lime is used to complete this process so that the phosphorous can precipitate and settle for removal. The most common is aluminum sulfate. The Chemical equations for removal are:

$$Al_2(SO_4)_3 + 2PO_4 \rightarrow 2AlPO_4 + 3SO_4$$

$$FeCl_3 + PO_4 \rightarrow FePO_4 + 3Cl$$

I. Solids treatment, handling, and disposal

Mixed Liquor Suspended Solids (MLSS) is the concentration of bacteria, solids, and any other undesirable material in sludge. To remove sludge, the MLSS is considered food for the activated microorganisms in the aeration process. It is often important to determine the food to microorganism ratio from the equation below:

$$F:M = \frac{S_{o,mg/L} Q_{o,MGD}}{V_{o,MG} X_{mg/L}}$$

$S_{o,mg/L}$ = Incoming BOD$_5$
$Q_{o,MGD}$ = Wastewater Flow Rate
$V_{o,MG}$ = Volume of aeration tank
$X_{mg/L}$ = Mixed Liquor Volatile Suspended Solids (MLVSS)

The time in which the sludge is held in the tank to allow the process to occur is the mean cell residence time:

$$\theta = \frac{VX}{Q_e X_e + Q_w X_w}$$

V = Volume of tank (ft^3)
Q_e = Effluent flow rate (ft^3/s)
Q_w = Waste flow rate (ft^3/s)
X = Concentration

J. Digestion

Digestion is a process of treating sludge that is too thick or bulky to be easily worked with for disposal. In other words, if the sludge is too thick it can be further broken down by digestion so that it can be moved more easily. There are 2 processes of digestion: aerobic and anerobic.

Aerobic digestion is putting the sludge in a large open holding tank for a period of time. In this tank the sludge is stirred and left open to air. This allows bacteria to consume the sludge reducing the solids. Often, up to 70% of the solids can be removed through this process.

Anaerobic digestion as the name suggests occurs without the use of oxygen. This process is more delicate in nature and proper care must be taken during as to not upset the desired result. However, it is often a more economical solution. Bacteria which does not require oxygen is introduced to the system. These bacteria, in a three-stage process, convert the sludge to gases which can then be released.

K. Disinfection

Disinfectants are chemicals which are used to kill bacteria that is present in water. In general when disinfectants are discussed, the chemical referred to is chlorine. Chlorine is easily the most widely used mainly because of the cost comparative to other types of disinfectants.

Chlorine however is a toxic substance and can be extremely dangerous to public health. It must be handled safely and properly.

In wastewater chlorine can be used to destroy common bacteria such as coliform. This is the presence of fecal matter in water supply.

L. Advanced treatment (e.g., physical, chemical, and biological processes)

Advanced treatment also known as tertiary treatment and is a final level of the wastewater treatment process. This phase handles any remaining pollutants that are still above allowable levels that have not been removed during the previous stages. Here are some of the pollutants that may be removed during this level.

Suspended Solids – At this point any solids remaining are very small in size and would need to be removed by more advanced techniques. This may involve microstrainers or filter beds which are able to remove very high gradation solids.

Phosphorous – This stage may require the removal of phosphorous. This is done through the use of chemical precipitation. This process utilizes aluminum, iron, and lime coagulates.

Ammonia – There are many processes for use of removal of ammonia to acceptable levels. These may include stripping, biological denitrification, breakpoint chlorination, anion exchange, and algae ponds.

VII. Water Quality

A. Stream degradation

Stream Degradation is the wearing away and lowering of a stream bed over time due to erosion from the flow of water within a stream. Often flow rates and velocities which are too high can cause this. The removal of the soil can affect the water quality by increasing the amount of sediment in the water and therefore decreasing clarity and oxygen in the stream.

B. Oxygen dynamics

Oxygen dynamics is related to the amount of oxygen present in flowing water. More specifically this type of oxygen is referred to as Dissolved Oxygen (DO). Simply put dissolved oxygen is the amount of gaseous oxygen present in a moving volume of water. DO is also affected by the temperature and must be adjusted appropriately

Dissolved Oxygen is one of the more relevant parameters when determining the quality of a mass of water so it is important to be able to understand and analyze its content. Typical concentrations of DO may vary greatly throughout the year for a given body of water. Concentrations may range anywhere from 1 mg/L to 20 mg/L. The level that is appropriate is based on the type of life that needs to be sustained. Certain organisms require a larger content of DO to survive than others. For example, larger marine animals such as trout or salmon may require 12-14 mg/L whereas others like pike may only need 3-4 mg/L.

Re-aeration is a process used to increase the Dissolved Oxygen content of a specific water source. This is where oxygen from the air is dissolved into the water by causing turbulence in the water. Turbulence causes the water to move rapidly by forces such as a physical force, wind or currents. This causes a mixing of the water and air in which increased amounts of oxygen will begin to dissolve into the body of water. Turbulence can be used to mix bottom and top areas of water which may have uneven amounts of dissolved oxygen. Often the top portion of water will have higher concentrations. Mixing them will more evenly distribute the dissolved oxygen. To determine the final DO concentration resulting from the mixing of two sources, use the following:

$$C_f = \frac{C_1 Q_1 + C_2 Q_2}{Q_1 + Q_2}$$

C = Concentration
Q = Flow rate

C. Total maximum daily load (TMDL) (e.g., nutrient contamination, DO, load allocation)

Total Maximum Daily Load often referred to as TMDL, relates to the Clean Water Act of 1972 which set the standards for pollutants in water. TMDL in accordance with this legislation provides the maximum amount of a pollutant which a body of water can receive that will not violate the quality standards. Some of the pollutants which are of concern are nitrogen, phosphorous, and sediment among others.

TMDL is the sum of all the pollutants entering a system and the inclusion of a factor of safety:

$$TMDL = WLA + LA + MOS + SV$$

WLA = Waste Land Allocation (Direct flow into the body such as pipes and ditches)
LA = Land Allocation (Pollutants from land areas)
MOS = Margin of Safety
SV = Seasonal Variation

D. Biological contaminants

Biological contaminants refers to the amount of organisms in the water. These organisms are also sometimes referred to as microbes. The microbes, because they are living, will reproduce if there is a sufficient supply of food. The food is called the substrate and may or may not be limited to facilitate the biological growth. The Monod equation is used to determine the rate at which substrate is converted into biomass which is simply the total mass of microorganisms in a given volume of water. The equation is as follows:

$$r_g = \frac{\mu_m X S}{K_s + S}$$

r_g = Rate of growth = $\frac{dX}{dt}$
μ_m = Maximum specific growth rate coefficient (time^{-1})
X = Concentration of microorganisms (mg/L)
S = Concentration of the growth limiting nutrient (mg/L)
K_S = Half-velocity coefficient (mg/L)

E. Chemical contaminants, including bioaccumulation

Chemical contaminants are a severe concern in water as they may impose health risks to the public. Water should be tested regularly for the presence of such chemicals and action taken immediately. Since chemicals pose a risk to human life, acceptable levels of risk need to be identified and associated with the concentrations of the chemical. The following equation can be used:

$$Risk = \frac{(Concentration)(Intake)(Absorption\ Factor)(Exposure)(Risk\ Factor)}{(Body\ Weight)(Lifetime)}$$

VIII. Drinking Water Distribution and Treatment

A. Drinking water distribution systems

As the name suggests, systems are developed so that drinking water can be safely and efficiently distributed to the populations. These systems may consist of many components such as pipes, reservoirs, pumps, storage tanks and many others. These components carry water from a centralized distribution plant which maintains regulated levels of safe drinking water.

B. Drinking water treatment processes

There is a large number of processes which can be performed to get water meeting quality standards. The selection of which processes are performed depends heavily on the characteristics of the water specific to a certain plant. The procedure can be divided into 3 components: Pretreatment, Treatment, and Special Treatment. Here we will provide a breakdown of what may be involved in each portion depending on the type of water that needs to be treated.

Pretreatment

Screening: As the name suggests suspended solids which are large enough to be physically removed by allowing water to flow through fine screens is an initial process that is necessary to remove any debris.

Microstraining: A second level of screening used to remove the finer debris. This process is very effective in the removal of algae.

Plain Settling: A removal of sediment by allowing the water to sit and the natural movement of sediments to fall to the bottom to occur.

Aeration: The rapid moving of water to allow mixing or the infusion of oxygen into the water. Aeration can have many benefits depending on the desired result. It can increase dissolved oxygen, decrease dissolved gases, reduce iron and manganese, or decrease odor and taste compounds.

Treatment

Lime Softening: As the name suggests this is the process of adding lime water (calcium hydroxide) to soften water. This additive will react with the calcium and manganese to form precipitates.

Coagulation and Sedimentation: This process is the addition of chemicals, called coagulates, to form together contaminants into solids which can then be removed. Coagulates form together precipitate which is called floc. This process is essential to the treatment of water and is covered in greater detail later on.

Rapid Sand Filtration/Pressure Sand Filtration: See section on filtration.

C. Demands

Water demands need to be measured and analyzed so that distribution systems may be properly designed. Water demand is most often specified as gallons per capita per day (gpcd). It can also be expressed as Average Annual Daily Flow (AADF) which as the name suggests is the average daily use of water per person averaged over a year time period. A common value used for basic design purposes is often taken as 165 gpcd but should be adjusted based on the intended water use whether it be residential, commercial, or industrial.

Besides the average flow demand throughout a day, there may be increased demands instantaneously which systems must have adequate capacity for. The average annual daily flow times a specified multiplier is often used to determine the instantaneous demand:

$$Q_{Instant} = M(AADF)$$

It is also important to note that per capita demand needs to account for the entire population but it must often be specified at what time period. Because of growth, a distribution system should meet some future predicted growth of population.

D. Storage

Water supplies need to be stored for a variety of uses and as well as to ensure adequate supply in times of growth or emergency. Water can be distributed from storage either through gravity or pumping. Gravity is available when there is a sufficiently high point in elevation relative to the population. Otherwise pumping is necessary. Water is most often stored in surface or elevated tanks. Within these tanks the elevation of the surface water is monitored to determine the appropriate distribution pressure. These are often monitored by altitude valves.

E. Sedimentation

A plain sedimentation tank is used to allow water which includes suspended sediments to settle out. The time and velocity for the particles to settle is a function of the temperature of the water, the particle size and the specific gravity of the particles (however this is often taken as 2.65 for analysis). Assumed settling velocities can be taken as the following to calculate the actual settling velocities:

Gravel: 3.28 ft/s
Coarse Sand: 0.328 ft/s
Fine Sand: 0.0328 ft/s
Silt: 0.000328 ft/s

Then the approach to determining the settlement time can be determined by first calculating the Reynolds number:

$$Re = \frac{v_s D}{\nu}$$

If Re < 1, use Stokes' Law:

$$v_s = \frac{(SG_{particle} - 1)D_{ft}^2 g}{18v}$$

If 1 < Re < 2000, the velocity is determined graphically.

If Re > 2000 use Newton's first law of motion as shown below:

$$v_s = \sqrt{\frac{4gD(SG_{particle} - 1)}{3C_D}} \text{ where } C_D = 24/Re$$

The time it takes to settle is then a function of the tank depth:

$$t_s = \frac{h}{v_s}$$

To ensure all particles have settled the settlement time must be less than the detention time (the total time the water stays in the settlement tank:

$$t_d = \frac{Ah}{Q}$$

F. Taste and odor control

There are many processes which can aid in the elimination of undesirable taste and odor in water. Some include chlorination, aeration and micro straining. To identify the presence of taste or order, the threshold odor number (TON) is established and can be calculated as per below:

$$TON = \frac{V_{Raw\ Sample} + V_{Dilution\ Water}}{V_{Raw\ Sample}}$$

Typically, a TON of less than 6 is desirable

G/H. Rapid mixing (e.g., coagulation)/Flocculation

This process as mentioned above is the addition of chemicals, called coagulates, to form together contaminants into solids which can then be removed. Coagulates form together precipitate which is called floc. For this reason we have combined two of the NCEES syllabus items since it is most appropriate to discuss these topics together. The most common type of coagulates are aluminum sulfate commonly referred to simply as alum. Others include ferrous sulfate and chlorinated copperas. Alum is often provided in doses in the range of 5-50 mg/L. There are three requirements for Alum to be effective:

1. A large enough quantity of Alum must be present to neutralize the negative particles present in the water
2. Enough alkalinity must be present to facilitate the reaction of aluminum sulfate to aluminum hydroxide
3. The PH must be within the acceptable range which is a function of the type on contaminant. Typically it is taken between 6-7

The amount of coagulate to successfully form floc must be determined. The equation for the feed rate is:

$$F_{lb/day} = \frac{D_{mg/L} Q_{MGD} \left(8.345 \frac{lb-L}{mg-MG}\right)}{PG}$$

F = Feed Rate
D = Dose
Q = Flow Rate
P = Purity
G = Availability (1.0 is not specified)

I. Filtration

Filtration is used to remove excess floc, precipitates from softening, algae, debris and any other suspended byproducts remaining in the treated water. The most common is rapid sand filtration. Rapid Sand Filtration is the filtering of water through a bed of sand and gravel as a medium for removing suspended particles. Water moves through a layer of sand in which the suspended particles will be held back by the sand. Depending on the type of filter, the loading rate can be anywhere from 2 – 10 gpm/ft². The loading rate can be determined by the following equation:

$$Load\ Rate = \frac{Flow\ Rate}{Area}$$

Filters often need to be cleaned and therefore there is a high maintenance cost. The pores between the filters will become clogged and need to get washed out. To counteract this is a process called back washing. This is where water is pumped slowly in the reverse direction of the water to be filtered so that the pores in the sand can be expanded to release any trapped material. It is important during backwashing to monitor the rise rate of the water to ensure it does not exceed the settling velocity of the smallest particle intended to be left in the filter. These rates are often taken as about 1-3 ft/min. The amount of backwash needed can be determined by:

$$V = A_{Filter}(Rise\ Rate)t_{Backwash}$$

J. Disinfection, including disinfection byproducts

Disinfectants were defined earlier in the wastewater section. Here we will discuss by products.

Chlorine in water produces the following chemical reaction depending on PH

PH > 4: $Cl_2 + H_2O \rightarrow HCl + HOCl$

PH > 9: $HOCl \rightarrow H^+ + OCl^-$

HCl and HOCL are hydrochloric and hypochlorous acids respectively. You can see that at PH greater than 9, the hypochlorous acid becomes hydrogen and hypochlorite ions.

K. Hardness and softening

Hardness is a measure of the presence of calcium and magnesium ions expressed as calcium carbonate ($CaCO_3$). Practically, hardness in water does not provide any health concerns but does have an effect on the usefulness of the water. One of the main concerns is often that hardness in water will greatly reduce the effectiveness of soap. It also has a detrimental effect on the pipes and storage facilities of a water distribution system.

There are two types of hardness:

Carbonate Hardness: Water containing Bicarbonate (HCO_3^-)

Noncarbonate Hardness: Remaining hardness not carbonate due to sulfates, chlorides, and nitrates.

Hardness can also be expressed as total hardness which is the sum of carbonate and noncarbonate hardness in mg/L as $CaCO_3$. There is a clear connection between the alkalinity of water and the hardness. The following assumptions can be made:

- If Total Hardness = Alkalinity, all hardness is carbonate and there are no sulfates, chlorides, or nitrates present
- If Total Hardness > Alkalinity, noncarbonate hardness is present
- If Total Hardness < Alkalinity, all hardness is carbonate and the remainder of the bicarbonate is from additional sources

Water softening is the removal of hardness through the use of lime and soda ash in mg/L as $CaCO_3$. It is important to note that lime will attack any carbon dioxide in water first and then begin with the removal of any carbonate hardness before the noncarbonate.

IX. Engineering Economics Analysis

A. Economic analysis (e.g., present worth, lifecycle costs, comparison of alternatives)

The PE exam will potentially provide examples of engineering economics which are geared towards assets in water resources. For these questions refer to the morning session equations for engineering economics.

MORNING BREADTH PRACTICE EXAMS

Question M1

The chart below gives an estimation of the area of sub-grade to be cut at stations along the baseline of a roadway. Determine most nearly the total volume of excavation in cubic yards from the data.

STATION	AREA OF CROSS SECTION (FT^2)
1+00	0
1+50	155
2+00	170
2+50	65
3+00	0

(A) 722
(B) 19500
(C) 800
(D) 19222

Question M2

Given the dimensions of the 8' tall proposed concrete retaining wall shown below determine the area of formwork in square feet required to complete construction of the wall portion only after the footing has been poured.

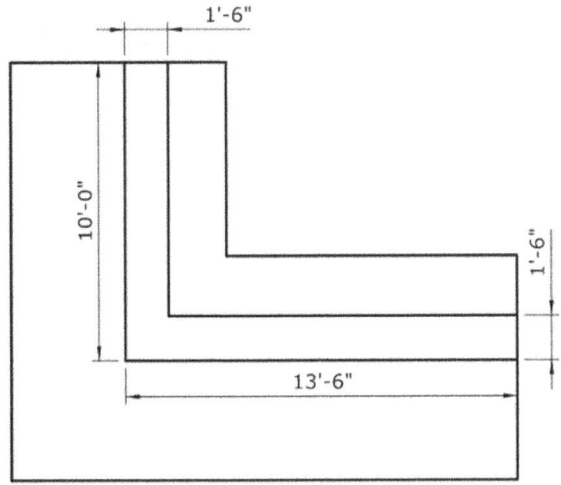

(A) 950
(B) 400
(C) 376
(D) 450

Question M3

Given the scheduling data shown below, Identify all of the tasks on the critical path.

Activity	Predecessor	Duration
A		5 (Months)
B	A	3
C	A	2
D	C	3
E	D	4
F	E	1
G	D	2
H	G	4

(A) A
(B) B
(C) C
(D) D
(E) E
(F) F
(G) G
(H) H

Question M4

Design alternatives are being proposed for a bridge replacement with a life span of 30 years. Alternative A has an initial cost of $100,000. It will also require anticipated maintenance costs of 10,000 and 15,000 at years 10 and 20 respectively. Use a rate of 3% and determine most nearly the present worth of the design alternative.

(A) $99050
(B) $110444
(C) $113250
(D) $115746

Question M5

A soil sample has 20% fines and over 50% finer than the no. 4 sieve. It also has a Liquid Limit of 55 and a Plastic Limit of 23. Determine the USCS classification group symbol of this soil.

(A) SC
(B) GC
(C) CL
(D) SM

Question M6

Given the beam and loading conditions shown below, determine most nearly the maximum service moment (kip-ft).

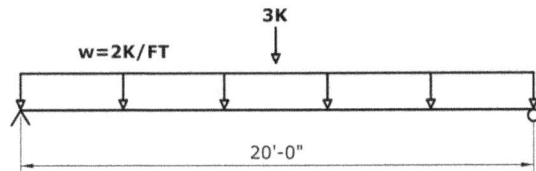

(A) 100
(B) 115
(C) 9.58
(D) 57.5

Question M7

Assign the appropriate theoretical effective length factor for the columns with end conditions as shown:

| 0.5 |
| 0.7 |
| 1.0 |
| 2.0 |

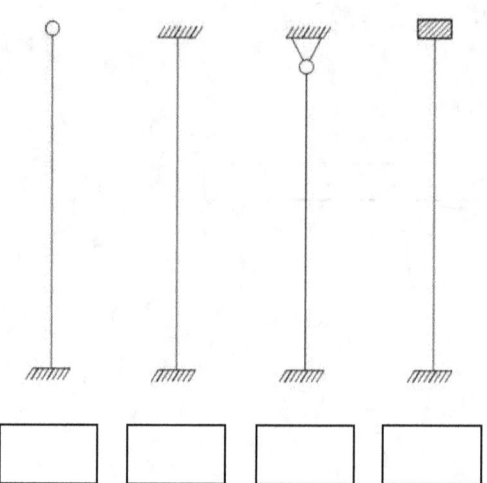

Question M8

Which of the following accurately represents the shear diagram for the beam and loading conditions shown?

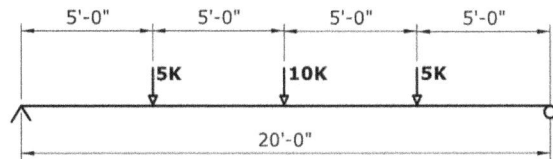

(A)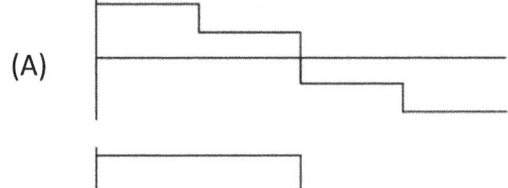

(B)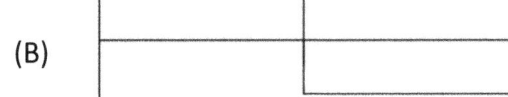

(C)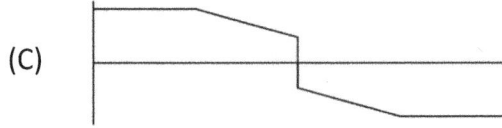

(D)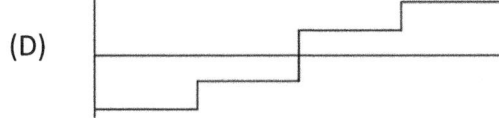

Question M9

Given the beam and loading condition shown below, determine most nearly the moment at point C in Kip-ft.

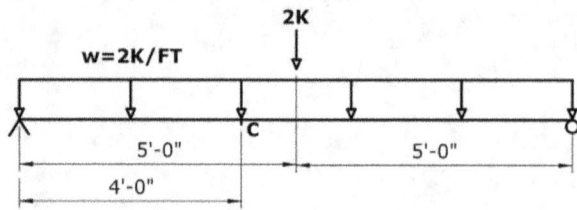

(A) -10
(B) 15
(C) 28
(D) 40

Question M10

Three 6-inch diameter x 12-inch-long concrete cylinders were taken from a concrete batch used to form a retaining wall. The cylinders broke at compressive loads of 102000, 111000, and 100500 pounds. Determine most nearly the average compressive strength of the concrete (psi).

(A) 3200
(B) 4000
(C) 3600
(D) 3700

Question M11

A horizontal curve has a point of curvature at station 1+50 and a point of tangency at station 2+25. If the interior angle of the curve is 10°15', determine most nearly the radius of the curve.

(A) 330'
(B) 420'
(C) 500'
(D) 600'

Question M12

A crest vertical curve has G_1 = 2%, G_2 = -4% and a PVI elevation of 155.0' at station 10+00. A bridge has a low point elevation of 170' to the bottom of the girder at station 8+00. Determine most nearly the vertical clearance between the road and low point of the bridge if the length of curve is 800'.

(A) 17.0'
(B) 20.5'
(C) 26.0'
(D) 29.5'

Question M13

A vertical Curve has G_1=-3% and G_2=2%. The PVI elevation is 150.0' at station 10+00. The Length of Curve is 600'. The station of the low point is _____.

Question M14

A horizontal curve is proposed to be constructed around a building. A car traveling along the curve has a line of sight past the building to an object in the road from the beginning of curve to the end of curve. Determine most nearly the minimum distance from the center of road to the building for a required stopping sight distance of 220' to the object if the radius of the curve is 700'.

(A) 8.63'
(B) 15.55'
(C) 20.20'
(D) 35.0'

Question M15

Backfill material is to be transported to a construction site to achieve a proposed grade behind a retaining wall. A total of 100 cubic yards is necessary in the final condition. Determine most nearly in cubic yards how much fill is needed to be taken from the offsite location if there is a swell factor of 1.07, an assumed loss during transport of 5%, and the backfill will be compacted to 90%?

(A) 100
(B) 102
(C) 106
(D) 109

Question M16

A mix design for a parking garage will use 500 lbs of cement per cubic yard. What volume of water in cubic feet per cubic yard should be used to achieve a w/c ratio of 0.65?

(A) 5.21
(B) 6.00
(C) 4.31
(D) 7.50

Question M17

A retaining wall without a batter is 15.5' tall from the bottom of footing and is used to hold granular soils with an angle of internal friction of 25. Determine most nearly the resultant service moment at the base of the footing using Rankine active earth pressure from the soil only. The angle of fill is horizontal. Neglect friction between the wall and the soil and use a soil density of 115 pcf.

(A) 24.7 k-ft
(B) 26.5 k-ft
(C) 28.9 k-ft
(D) 32.1 k-ft

Question M18

A saturated soil sample has a weight of 50 lbs. The sample is placed in an oven and then weighed to measure 42 lbs. The specific gravity of the soil was determined to be 2.4. Determine most nearly the void ratio of the sample.

(A) 0.28
(B) 0.40
(C) 0.42
(D) 0.46

Question M19

A vehicle is traveling at a velocity of 100 ft/s on a 2% incline. If the driver has a 2 second breaking perception reaction time, determine the total distance in feet it takes to stop the vehicle. Assume a coefficient of friction of 0.3.

(A) 200.0
(B) 684.5
(C) 480.9
(D) 700.0

Question M20

Which of the following appropriately defines a design 100-year storm?

(A) A storm that occurs once per year over a span of 100 years
(B) A storm event with the 100th greatest intensity in a given year
(C) A storm with an intensity which will only occur once in 100 years
(D) A storm with an intensity of 100

Question M21

Assuming fully turbulent flow, determine most nearly the velocity in ft/sec for a pipe with a diameter of 2 ft, a length of 500 ft and a head loss due to friction of 3 ft. Use a Darcy friction factor 0.02.

(A) 1.0
(B) 5.0
(C) 6.2
(D) 11.4

Question M22

A rectangular open channel is 6 ft wide and has a water depth of 3 ft. Determine most nearly the flow rate (cfs). The channel has a roughness coefficient of 0.015 and a slope of 0.003.

(A) 130
(B) 138
(C) 145
(D) 200

Question M23

For a head loss due to friction of 40 ft, determine most nearly the volumetric flow rate in gallons per min for a pipe with length 100' and diameter of 0.25 ft. Use a Hazen-Williams coefficient of 140.

(A) 200
(B) 265
(C) 410
(D) 432

Question M24

Which of the following defines the time base for a stream hydrograph?

(A) Time from the base flow until the peak flow
(B) Time from the peak flow until flow drops below the base flow
(C) Time that the flow exceeds the base flow
(D) Time that the flow drops below the base flow

Question M25

A circular culvert is to be designed to handle the runoff from two areas of drainage. The first has an area of 15 Acres, a runoff coefficient of 0.18, and a rainfall intensity of 1.5 in/hr. The second has an area of 10 Acres, a runoff coefficient of 0.22, and a rainfall intensity of 1.5 in/hr as well. Determine most nearly the minimum culvert area in square feet to limit the flow velocity to 0.26 ft/s.

(A) 28.28
(B) 29.50
(C) 31.0
(D) 62.0

Question M26

Given the truss configuration shown below, identify the 0-Force members. The truss is simply supported and points A and H. (point and click example)

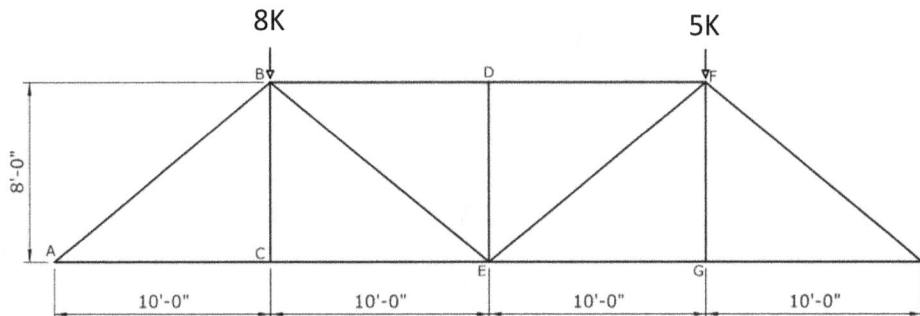

Question M27

Given the truss configuration shown below, what is the support reaction at node A? Neglect the self-weight of the truss. The truss is simply supported and points A and H.

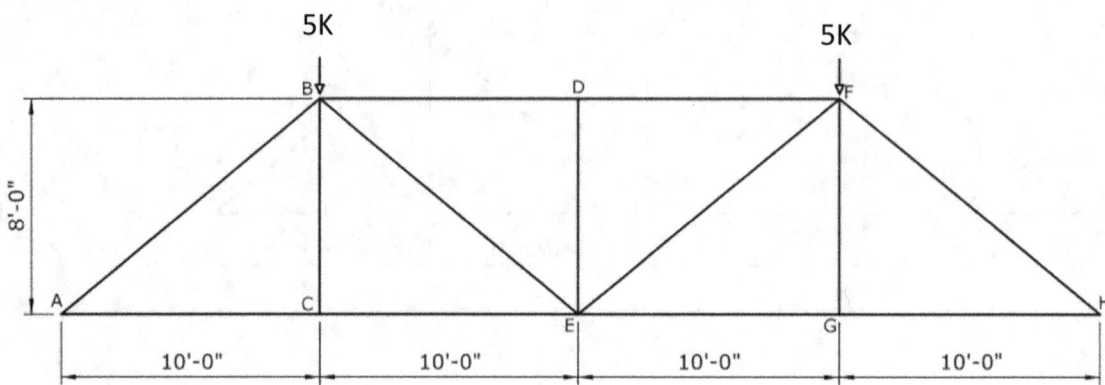

(A) 2 Kips
(B) 5 Kips
(C) 6.67 Kips
(D) 10 Kips

Question M28

Given the truss and loading condition in question M27, what is the axial force in member C-E? Neglect the self-weight of the truss.

(A) 0 Kips
(B) 5 Kips
(C) 6.24 Kips
(D) 8.71 Kips

Question M29

Which of the following is most often the cause of rebar corrosion?

(A) Chloride intrusion of concrete
(B) Freeze-thaw action
(C) Delayed Ettringite Formation
(D) High water to cement ratio

Question M30

Which of the following mix design properties has the greatest impact on the strength of the concrete?

(A) w/c ratio
(B) Percent Coarse Aggregate
(C) Void Ratio
(D) Percent Fine Aggregate

Question M31

Water exits a reservoir through a pipe 50' below the water surface. Determine the velocity of flow at the exit of the pipe. Assume frictionless flow and the discharge is at atmospheric pressure.

(A) 45.66 ft/s
(B) 56.75 ft/s
(C) 60.10 ft/s
(D) 75.55 ft/s

Question M32

Which of the following is not an assumption of the Bernoulli energy conservation equation?

(A) The fluid is incompressible
(B) There is no fluid friction
(C) Changes in thermal energy are negligible
(D) The potential energy is zero

Question M33

Flow from pipes A and B connect to exit out a single pipe C and have velocities of 1.2 and 0.8 ft/s respectively. If pipe A has an area of 3.0 ft² and pipe B an area of 4.0 ft², what is the flow rate for pipe C in cubic feet per second?

(A) 3.6
(B) 4.0
(C) 6.0
(D) 6.8

Question M34

After an onsite investigation, the soil profile below has been developed from borings. Determine most nearly the effective stress at a depth of 40 ft for the saturated soil densities indicated.

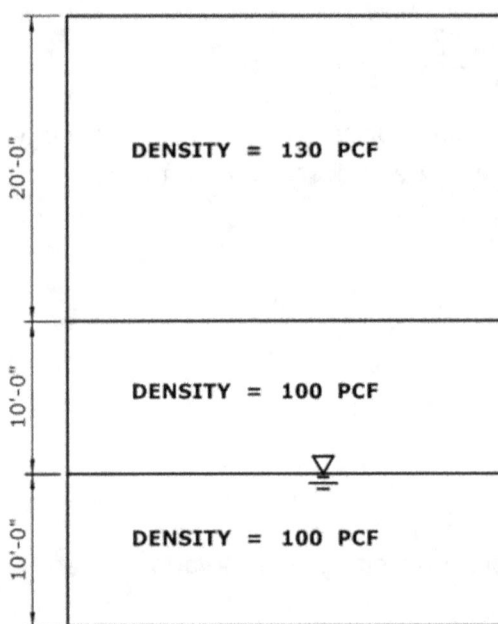

(A) 2600 psf
(B) 3550 psf
(C) 3975 psf
(D) 4600 psf

Question M35

A simply supported 4" wide X 6" deep beam is 15' long and has two point loads of 10 kips applied at 5' and 10' from the left end, what is the maximum bending stress of the cross section?

(A) 25 ksi
(B) 35 ksi
(C) 50 ksi
(D) 55.5 ksi

Question M36

A retaining wall is to be designed by limiting the shear stress in the stem to 0.1 ksi/ft. As shown below the horizontal force due to earth pressure is calculated as 9 Kips. The retaining wall has a batter of 8:1. Determine most nearly the minimum width of the base of the stem, b, if the depth to the flexural reinforcement is 6".

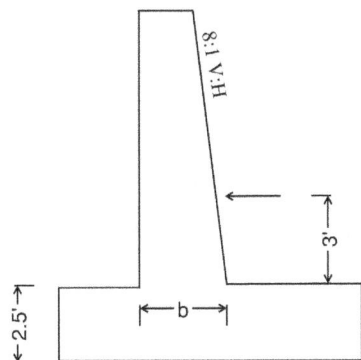

(A) 6.0"
(B) 7.5"
(C) 8.25"
(D) 12"

Question M37

Which of the following roadside safety barriers is most appropriate for an object which is 2'-6" perpendicular from the edge of roadway?

(A) Concrete Barrier
(B) 3-Cable Guiderail System
(C) Metal Beam Rail
(D) None

Question M38

For the traffic counts shown below, determine the peak hourly traffic volume.

Time	Volume
8:00-8:15	500
8:15-8:30	560
8:30-8:45	650
8:45-9:00	625
9:00-9:15	630
9:15-9:30	600
9:30-9:45	540
9:45-10:00	460

(A) 2260
(B) 2455
(C) 2505
(D) 2600

Question M39

Given a 6" wide X 12" deep cantilever beam with a point load at the free end of 1 K, determine the maximum deflection on the beam for a length of 10'. Use a modulus of elasticity of 3605 ksi.

(A) 0.10"
(B) 0.18"
(C) 0.28"
(D) 0.50"

Question M40

Determine the specific gravity of a soil sample. The total volume is 2 cubic ft and the volume of the soil is 1.5 cubic ft. The degree a saturation is 75% and the moisture content is 0.1.

(A) 2.2
(B) 2.5
(C) 2.8
(D) 2.95

Question M41

A soil sample is analyzed in the lab and it is determined to have 32% fines. The Liquid Limit is 36 and the Plastic Limit is 24. Determine the AASHTO Soil Classification of the sample.

(A) A-1-a
(B) A-2-4
(C) A-2-6
(D) A-4

Question M42

Charpy V-Notch is a test for which property of steel?

(A) Durability
(B) Strength
(C) Ductility
(D) Toughness

Question M43

Loads are determined for the design of a third story building. The types and magnitude of the loads are as follows:

Floor Slab Self-Weight = 100 psf
Flooring = 10 psf
Pipes and Utilities = 2 psf
Pedestrian Loading = 80 psf
Furniture = 40 psf

Using the Load combination of 1.2DL + 1.6LL, the design distributed load is _____.

Question M44

A concrete beam is 6" wide x 12" deep and has a uniform distributed load of 2 k/in as shown below. Determine most nearly the maximum deflection of the beam using a modulus of elasticity of 3605 ksi.

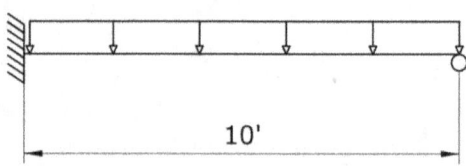

(A) 0.25"
(B) 0.55"
(C) 0.68"
(D) 0.72"

Question M45

A state department of transportation initiates a project to construct a maintenance facility for the housing of snow plows. The footprint of the building falls within state property however there is a need to store materials in an area past the property line of an adjacent privately-owned facility. Which of the following is the most appropriate Rights of Way action to complete construction?

(A) Temporary Easement
(B) Permanent Easement
(C) Property Acquisition
(D) No action required

Question M46

The cross section for a 20' long retaining wall is shown below. The dimensions of the cross section are consistent throughout the length. A chart is developed for the items identified for the completion of the work and unit prices have been developed using previous projects. Determine the cost of the work for the identified items.

Item	Unit	Unit Price
Concrete Wall	Cubic ft.	30
Backfill	Cubic Yd.	100
6" Dia. Drain	Linear ft.	20

(A) 15007
(B) 22104
(C) 33550
(D) 55250

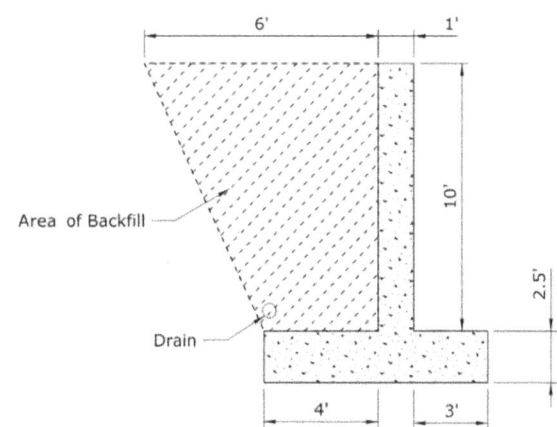

Question M47

A multi-use trail is to be constructed by crossing under an existing roadway using precast box culvert sections and detouring traffic. Which of the following is an appropriate sequence for the construction activities?

(A) Close roadway to traffic, place box culvert, backfill, pave road and open to traffic
(B) Set up detour signalization, close roadway to traffic, excavate, place box culvert, backfill, pave road and open to traffic
(C) Set up detour signalization, close roadway to traffic, excavate, place box culvert, allow for curing time, backfill, pave road and open to traffic
(D) Set up detour signalization, excavate, close roadway to traffic, place box culvert, backfill, pave road and open to traffic

Question M48

The chart below shows lift capacities (in pounds x 1000) of a crawler crane with a 142,000 lb counter weight. Using a boom length of 120' determine most nearly the maximum swing radius when lifting a structural member weighing 75 tons.

Boom(ft)	100	120	140
Radius (ft)			
24	191.7	175.9	157.9
30	160.2	152.2	140.6
34	136.0	134.7	127.5
40	107.8	107.9	107.7

(A) 24'
(B) 30'
(C) 34'
(D) 40'

Question M49

A bridge with approximately an 80' span is to be replaced in an urban area with high average daily traffic numbers. It has been determined that it is acceptable to detour the road for short period of time to allow for the replacement of the bridge. Which of the following is the most appropriate construction method?

(A) Cast in place reinforced concrete
(B) Precast concrete
(C) Prestressed-precast concrete
(D) Post-tensioned concrete

Question M50

A temporary shoring tower is being used to support a 30' long steel girder beam as shown below. Determine most nearly the minimum distance L which will limit the unfactored design reaction in the tower to a maximum of 5 kips. The steel beam has a self-weight of 288 lb/ft.

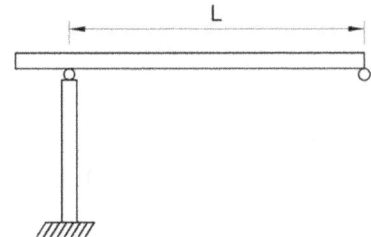

(A) 18'
(B) 22'
(C) 26'
(D) 29'

Question M51

Which of the following excavation scenarios would require the use of a temporary earth retaining systems?

(A) 8' deep cut into sandy soil
(B) 4' deep cut into clay
(C) 10' deep cut into stable rock
(D) 6' deep cut into stable rock

Question M52

An open cut excavation takes place to facilitate the placement of concrete pipes. The excavation is sloped to prevent failure at a rate of 1.5:1.0 V:H. A boring log indicates the soil is saturated clay with a density of 125 pcf and a cohesion of 300 psf. What is the maximum length of flat workspace for a factor of safety of 1.5 if the available right of way is 100' long? Use a slope stability number of 6.2.

(A) 20.0'
(B) 61.0'
(C) 73.6'
(D) 90.5'

Question M53

A pipe as shown below handles water with a flow rate of 2.0 cubic feet per second. Determine the minor head loss between point A and B. The K factor for a 90° bend was determined as 0.9.

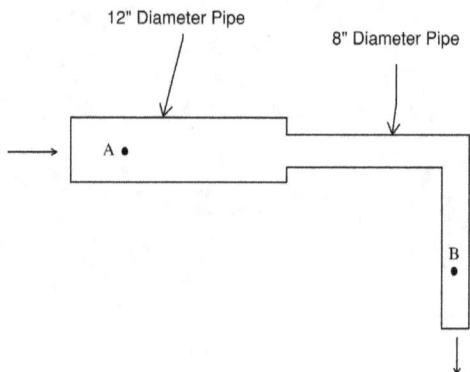

(A) 0.12'
(B) 0.46'
(C) 1.20'
(D) 2.29'

Question M54

A 25' long 3' x 3' concrete column is constructed to support a concentrated load of 50 kips. The column is fixed at the base and free to rotate but not translate at the top. Determine the Euler critical stress if the Modulus of Elasticity is 3605 ksi.

(A) 18.2 ksi
(B) 45.0 ksi
(C) 58.8 ksi
(D) 66.8 ksi

Question M55

Select all of the following USCS Group Symbol soil types that would be expected to undergo significant primary consolidation.

(A) GW
(B) GP
(C) SP
(D) CH
(E) CL
(F) GM
(G) SM
(H) Pt

Question M56

A horizontal curve is shown below. Label the appropriate elements of the superelevation.

- Tangent runout
- Superelevation Transition
- Superelevation Runoff

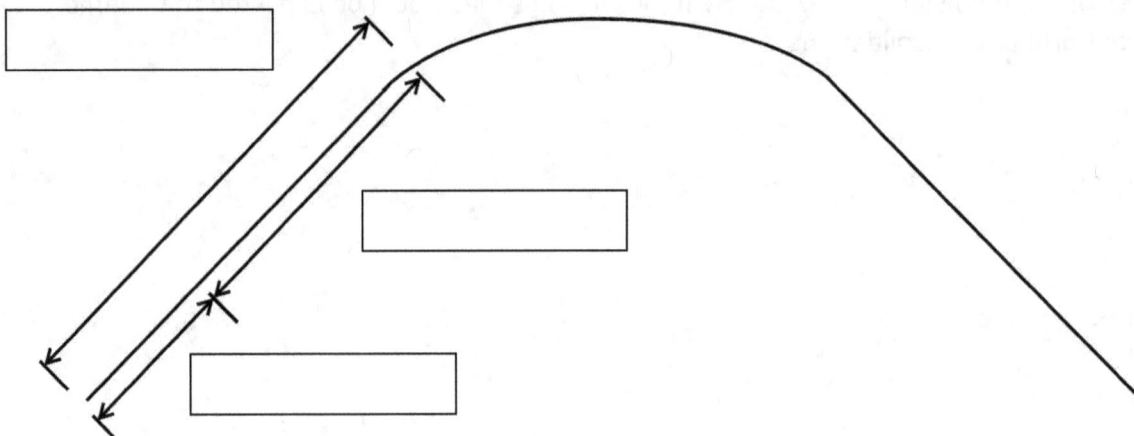

Question M57

A detention pond is to be designed to limit the flow of runoff at the base of a slope. Water exits the pond by a 2' diameter pipe and it is desired to limit the flow velocity to 0.25 feet per second. Determine most nearly the minimum size of the pond in cubic yards to prevent overtopping of the pond for a 6-hr storm event with an average runoff flow rate into the pond of 0.85 cubic feet per second.

(A) 25
(B) 52
(C) 75
(D) 150

Question M58

A soil sample is taken using a split spoon method with a 140 lb hammer. The chart below shows the number of blows for each 6" increment. Determine most nearly the average N value of the soil sample for the depth shown.

Depth (in)	Blows per 6" Increment
6	5
12	10
18	16
24	18
30	20
36	28
42	32

(A) 22
(B) 28
(C) 35
(D) 41

Question M59

A permeability test is performed on a cylindrical soil sample with an approximate diameter of 6" and a length of 12". It was measured that 1 cubic inch of water took 10 seconds to enter and exit the sample. The difference in pressure head was measured as 3". Determine most nearly the hydraulic conductivity in in/s of the sample.

(A) 0.005
(B) 0.014
(C) 0.110
(D) 1.140

Question M60

During the construction of a bridge replacement project bearing pads are surveyed to verify they are set at the appropriate elevation. The first measurement is recorded as 158.5682'. Which of the following is the appropriate recorded measurement with the appropriate level of accuracy?

(A) 158.568
(B) 158.57
(C) 158.6
(D) 159.0

Question M61

Which of the following methods is most often used to permanently control sediment from runoff of excavated areas?

(A) Silt Fence
(B) Hay Bales
(C) Seeding and Turf Establishment
(D) Erosion Control Fabric

Question M62

A 14" x 14" concrete column fixed at the base is subjected to an eccentric load of 5 kips as shown below. Determine most nearly the maximum compressive stress in the column at the base. Ignore self-weight of the column.

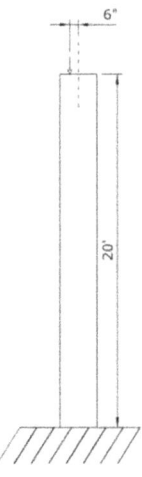

(A) 0.091 ksi
(B) 0.140 ksi
(C) 0.255 ksi
(D) 0.820 ksi

Question M63

A project is determined to need 14' x 10' box culverts of varying lengths in 3 different locations. Which of the following is the most appropriate unit of measure for the estimation of box culvert item?

(A) Linear ft.
(B) Square ft.
(C) Each
(D) Lump Sum

Question M64

A 10' X 10' spread footing is shown below, determine ultimate moment at the critical section for the design of flexural reinforcement. The factored point load includes the self-weight of the column and footing and is 50 K.

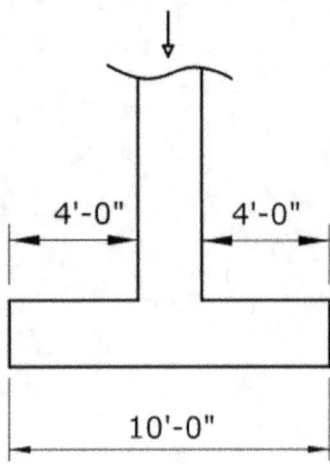

(A) 10 K - ft
(B) 18 K - ft
(C) 25 K - ft
(D) 40 K - ft

Question M65

Determine most nearly the time of concentration for sheet flow of a watershed which has a distance to the outlet of 250'. The watershed is short prairie grass which has a Manning's Roughness Coefficient of 0.15 and has a slope of 0.05 ft per ft. The 2 yr, 24 hr rainfall is given as 2.2 inches.

(A) 0.284 hrs.
(B) 0.313 hrs.
(C) 0.385 hrs.
(D) 0.486 hrs.

Question M66

Traffic data from an observation of highway is shown below. Determine most nearly the peak hour factor.

Time Interval	Volume(# of vehicles)
7:00-7:15	1900
7:15-7:30	2200
7:30-7:45	2350
7:45-8:00	2000
8:00-8:15	1700
8:15-8:30	1200

(A) 0.85
(B) 0.87
(C) 0.90
(D) 0.95

Question M67

A storm event produces an average depth of 3.5 inches. Using an NRCS curve number of 79, determine the direct runoff.

(A) 1.56"
(B) 1.78"
(C) 2.5"
(D) 4.4"

Question M68

A reinforced concrete wall is supported on a strip footing 8' wide and 2.5' deep from the soil surface to the bottom of the footing. The soil below is cohesion less, has a unit weight of 125 pcf, a density bearing capacity factor of 2.5 and a depth bearing capacity factor of 4.4. Determine the allowable bearing capacity of the soil for a safety factor of 2.5.

(A) 0.925 ksf
(B) 1.650 ksf
(C) 2.255 ksf
(D) 2.58 ksf

Question M69

Which of the following cement types is most appropriate for the construction of a mat foundation where sulfate attack is not a concern?

(A) Type I
(B) Type II
(C) Type III
(D) Type IV

Question M70

During a 3 hr storm event, a hydrograph is developed as shown below by taking measurements at 1-hour intervals of the flow rate. The storm intensities in half hour increments were also recorded and provided in the table below. Determine most nearly using a histographic approximation of the hydrograph the total volume of water in cubic feet during the lag time of the storm event.

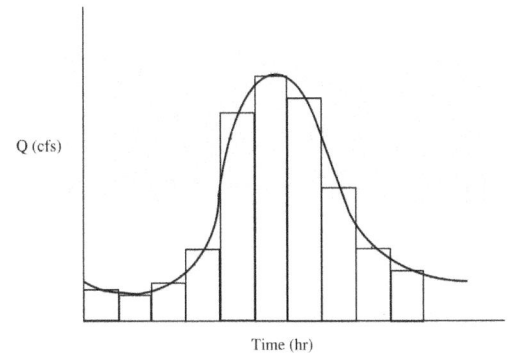

	A	B	C	D	E	F	G	H	I	J
Flow Rate (cfs)	1.1	0.5	1.4	2.5	6.6	8.6	6.9	4.5	2.6	1.6
Time (hrs)	1	2	3	4	5	6	7	8	9	10

Intensity (in/hr)	0.2	0.5	1.5	1.8	0.4	0.1
Time (hrs)	0.5	1	1.5	2.0	2.5	3.0

(A) 1857
(B) 54180
(C) 146160
(D) 211680

Question M71

An 8' wide x 8" deep precast wall panel is to be erected by being lifted from a flat position as shown below. The self-weight is 0.8 K/ft. Determine the minimum strength of concrete, f'_c, to prevent tensile cracking during lifting. Assume the wall panel can be analyzed as a simply supported beam between the pick points. Do not incorporate any safety factors or load factors.

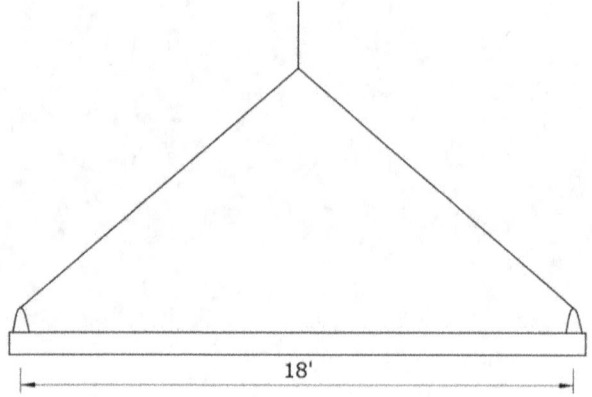

(A) 2250 psi
(B) 2480 psi
(C) 2570 psi
(D) 3200 psi

Question M72

A department of transportation is trying to determine its fiscal budget for the next five years. Three projects are being evaluated specifically which have a proposed cost of 100000, 175000, and 225000 and are scheduled to be paid for 3 years, 4 years, and 5 years from today respectively. Using an inflation of 3%, determine the total cost to the department.

(A) 475000
(B) 500000
(C) 525090
(D) 567073

Question M73

Which of the following is the least appropriate shape for the design of flexure about the X-axis?

(A)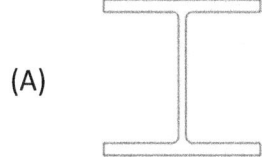

(B)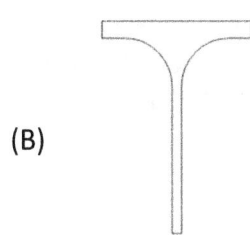

(C)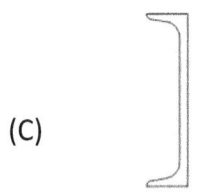

(D)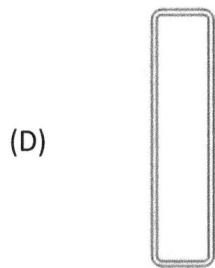

Question M74

Which of the following is not a method for the protection of steel against corrosion?

(A) Weathering Steel
(B) Galvanized Steel
(C) High Strength Steel
(D) Painting Systems

Question M75

A bridge is to be replaced which carries a local road over a highway. A boring log reveals the soil types for the construction of the north east wingwall of the bridge to be sandy soils to a depth of 10 ft. and then bedrock 10 ft. and below. The wingwalls are anticipated to be about 10' high. Which of the following is the most appropriate foundation type for the wingwall?

(A) Spread Footing
(B) Micro-pile
(C) Pile
(D) Drilled Shaft

Question M76

A project schedule is shown below with tasks designated by letters and durations above each task in days to complete for a single crew. This project includes two separate crews of workers and the crews can work either separately or simultaneously on any task at any time. The cost per hour for workers is $50/worker and crews are made up of 4 workers each. Work days are 8 hrs. long. Determine the minimum cost of labor to complete the project.

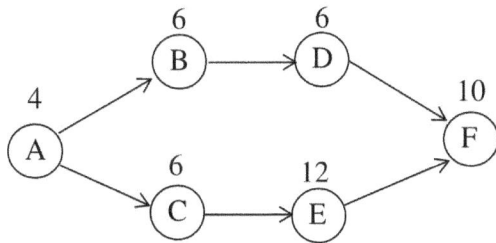

(A) 5250
(B) 22200
(C) 70400
(D) 100500

Question M77

A project schedule is determined and an activity diagram is developed as shown below. Predecessors are indicated with arrows and the associated original durations are indicated in days by the numbers. However, due to competing priorities, the durations of tasks C, D, and F need to increase by 1, 4, and 3 days respectively. Select the activities on the new critical path (point and click example).

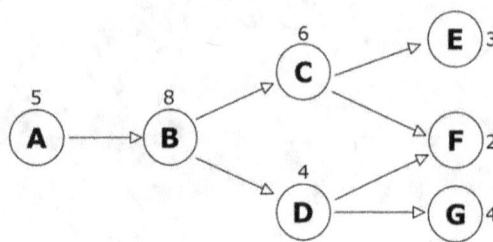

Question M78

An 8" thick one-way slab spans 20' between wall panels and is 10' wide. The slab is fixed at the connection to the walls. Including self-weight, the slab is to be designed for a distributed load of 0.5 ksf. Determine most nearly the un-factored design moment in the design of the slab.

(A) 9.50 k-ft
(B) 12.55 k-ft
(C) 14.20 k-ft
(D) 16.67 k-ft

Question M79

The roadway cross section below includes a construction stake with the markings shown. Determine the elevation of the top of curb if the height of the curb from the roadway is 10".

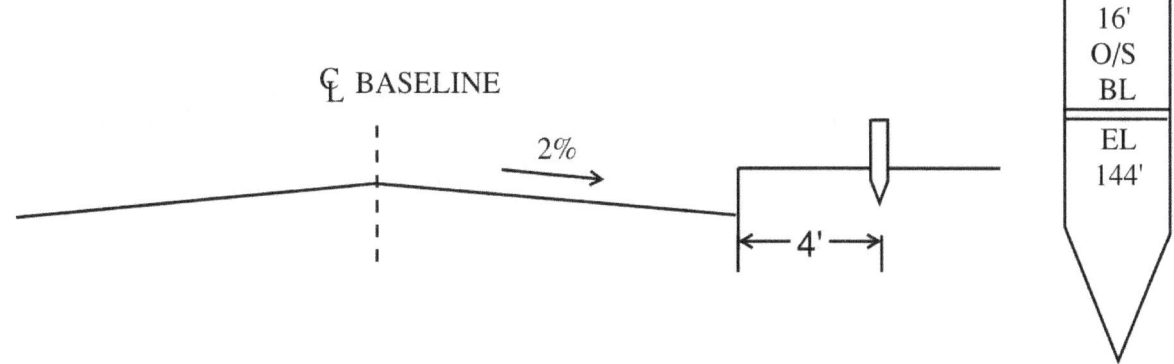

(A) 143.75'
(B) 144.0'
(C) 144.60'
(D) 145.8"

Question M80

A tensile lab test is performed on a cylindrical steel member having a diameter of 0.25". From the results below determine most nearly the yield strength (ksi) of the steel.

Load (lbs)	437	778	1502	2033	2262	2656	2428	2392	2601	3180	3297
Elongation (in.)	0.019	0.021	0.025	0.027	0.028	0.03	0.031	0.059	0.098	0.23	0.32

(A) 36
(B) 45
(C) 54
(D) 60

WATER RESOURCES AND ENVIRONMENTAL DEPTH PRACTICE EXAM

Question W1

Wastewater flows into a rectangular plain sedimentation clarifier at a rate of 1.2 MGD. The clarifier is 45' x 35'. It is determined that the detention time for the treatment of wastewater is 2 hrs. Determine the minimum depth of the clarifier if it is required to have a minimum of 6' above the calculated basin depth.

(A) 9.0'
(B) 14.5'
(C) 22.0'
(D) 35.6'

Question W2

Wastewater flows into a 1.57 MG aeration tank for sludge treatment at a rate of 5.0 MGD. The mixed liquor volatile suspended solids concentration is 1500 mg/L. The plant maintains a food to microorganism ratio of 0.2. If the BOD removal efficiency is 90%, determine the effluent BOD concentration.

(A) 9.42 mg/L
(B) 25.57 mg/L
(C) 35.8 mg/L
(D) 55.7 mg/L

Question W3

Determine the maximum concentration of Trichloroethylene (TCG) which can be in drinking water that will limit the cancer risk to 0.003. The representative person is 68 kg and will live for 75 years. The TCG is present in the subjects drinking water for 10 years at a rate of 1.2 liters per day and an estimated 95% absorption rate. TCG has a risk factor of 0.4 kg-day/mg.

(A) 1.0 mg/L
(B) 2.5 mg/L
(C) 3.36 mg/L
(D) 11.4 mg/L

Question W4

A network of two parallel pipes are shown in the schematic below with diameters, lengths and Hazen Williams coefficients as shown. If the velocity of the flow in pipe 1 is measured to be 6.0 ft/s, determine the flow rate (cfs) for the system.

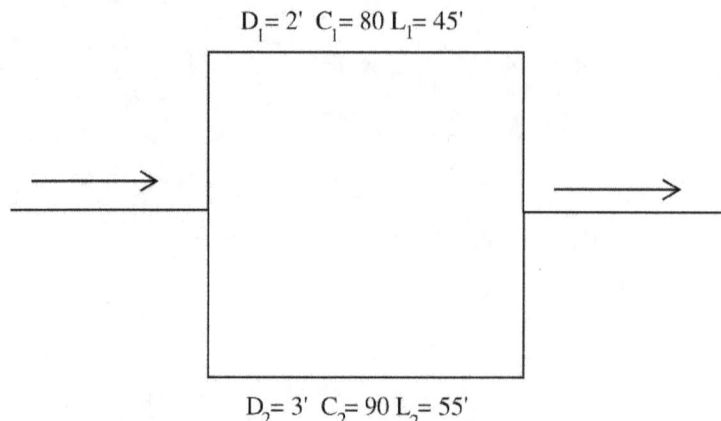

(A) 74
(B) 88
(C) 155
(D) 250

Question W5

An existing drainage pipe is 2' in diameter. It is determined there is a need to upgrade the drainage system to handle a 50% increase in flow rate, all of which will be handled by a supplemental pipe to be installed alongside the existing pipe. The velocity of the flow in the supplemental pipe will be equal to the existing pipe. A pitot static gauge is used to measure the flow of the existing pipe. A mercury manometer measures a 2" difference in elevation. The density of mercury is 848.6 lb/ft³. Determine the minimum area of the supplemental pipe in square feet.

(A) 1.57
(B) 2.22
(C) 5.75
(D) 12.66

Question W6

Six wastewater samples are taken each having a volume of 4 mL. The samples are put into bottles and are all diluted by adding 250 mL of water. Three of the samples are titrated immediately and three are titrated after a time of 5 days. The dissolved oxygen content is noted in the chart below. Determine most nearly the average ultimate BOD of the samples for a deoxygenation rate of 0.11 day^{-1}.

Incubated Samples

Sample #	DO Content (mg/L)
1	2.1
2	2.3
3	2.4

Initial Samples

Sample #	DO Content (mg/L)
4	6.2
5	6.8
6	6.4

(A) 221.45 mg/L
(B) 320.22mg/L
(C) 371.8 mg/L
(D) 390.56 mg/L

Question W7

A wastewater treatment plant has 5 lbs remaining of aluminum sulfate for phosphorous removal. Given the following characteristics, determine the amount, in pounds, of phosphorous which can be removed with the remaining quantity.

Molar Mass for Aluminum Sulfate = 342.15 g/mol
Molar Mass of PO_4 = 94.97 g/mol
Phosphorous Mass Percentage of PO_4 = 32.62%

(A) 0.91
(B) 2.2
(C) 5.6
(D) 20.0

Question W8

A drinking water supply is to undergo some coagulation using aluminum sulfate. The water treatment plant can provide a feed rate of 50 lb/day. The alum that is used has a purity of 45% and has no availability concerns. Determine most nearly the appropriate dose of the coagulate if the water has a flow rate 2.1 MGD.

(A) 1.28 mg/L
(B) 1.76 mg/L
(C) 2.20 mg/L
(D) 3.14 mg/L

Question W9

Six open channel concrete cross sections with dimensions and water elevations are shown below. Choose the one rectangular and the one trapezoid that has an ideally efficient section (point and click example).

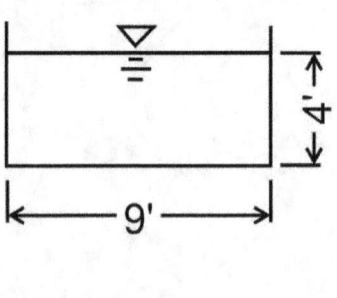
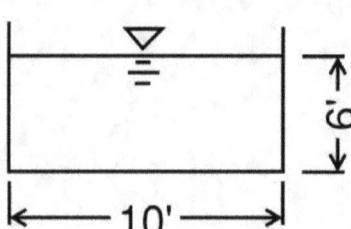
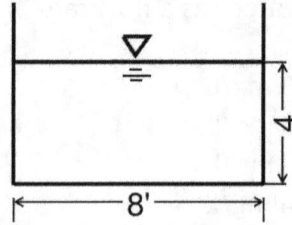
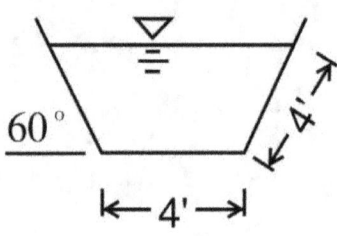
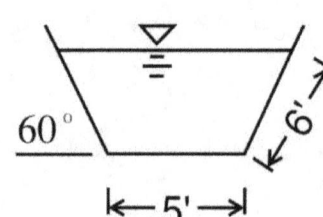
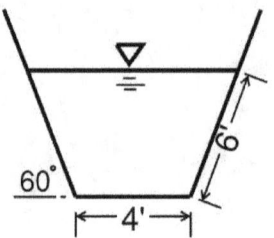

Question W10

A sludge processing aeration tank has a diameter of 20 ft. The target design mean cell residence time is 10 days. The influent has a MLSS of 2000 mg/L and flows at a rate of 1062 cubic ft/day. The flow out of the tank is divided into effluent and waste where 85% is effluent and 15% is waste. The effluent has a MLSS of 30 mg/L and the waste is 2200 mg/L. Determine the minimum required depth of the tank for the mean residence time.

(A) 6.0'
(B) 12.4'
(C) 18.6'
(D) 25.4'

Question W11

A hydraulic jump occurs in a 5' wide rectangular channel due to upstream subcritical flow as shown below. The flow rate at point A is 300 cubic feet per second at a depth of 4'. Determine most nearly the velocity of flow at point B.

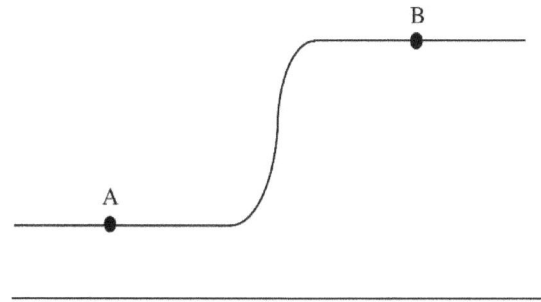

(A) 9.98 ft/s
(B) 10.45 ft/s
(C) 15.0 ft/s
(D) 22.65 ft/s

Question W12

A confined aquifer is 50' thick and has a well pumping from it. It is determined that the aquifer material has a permeability of 20 gal/day-ft². Observation wells are set up at a distance of 40' and 75' from the well and measurements show the water elevation at 20' and 15' deep respectively. Determine the discharge from the well in gallons per day.

(A) 9000
(B) 22860
(C) 35670
(D) 49977

Question W13

A small-town engineering department is evaluating the need for an increase in water supply to meet the anticipated future demand. Measurements indicate that the towns average annual daily flow per capita is 85 gallons per person per day. From the census, it is seen that the town population has grown on an average of 3% per year. If the population is currently 25,000 residents, determine the instantaneous demand necessary in gallons for a ten-year projection if the average flow per capita per day remains the same. It was determined to use a demand multiplier of 2.2.

(A) 1.02×10^6
(B) 2.21×10^6
(C) 5.55×10^6
(D) 6.28×10^6

Question W14

A total maximum daily load (TMDL) analysis is to be performed by the state DEEP for a river. All sources need to be identified to complete the analysis. Which of the following can be classified as a Waste Load Allocation (WLA)?

(A) Leakage identified from a deteriorated industrial septic tank
(B) Discharge from a local wastewater treatment plant
(C) Runoff from the farm upstream
(D) Fuel discharge from the local marina

Question W15

A dual media rapid sand filtration tank has been taken out of service for maintenance. The tank is to be backwashed for cleaning. It has a cross sectional area of 200 ft² and is to be filled to a depth of 10 ft. The sand in the tank has a Reynolds number of 0.68 and the diameter of the particles can be assumed to be 0.0002 ft. If the kinematic viscosity of the water at the current temperature is 9.62 x 10⁻⁶ ft²/s, determine most nearly the minimum time for the backwash operation in minutes. Use an SG = 2.65 and assume the area of filter is equal to the area of tank.

(A) 8.0
(B) 13.5
(C) 35.5
(D) 45.0

Question W16

A sample of water is determined to have a total hardness of 200 mg/L as CaCO₃ and a noncarbonate hardness of 50 mg/L as CaCO₃. There is also 20 mg/L as CaCO₃ of carbon dioxide present. The water is to be softened using lime with a purity of 92% and soda ash with a purity of 95%. Determine most nearly the appropriate amounts in mg/L as CaCO₃ of both the lime for the carbonate removal and the soda ash respectively for softening of the water.

(A) 20 Soda and 120 Lime
(B) 55 Soda and 150 Lime
(C) 48 Soda and 175 Lime
(D) 53 Soda and 185 Lime

Question W17

The cross section for a trapezoidal weir is shown below. Determine the discharge in cubic feet per second.

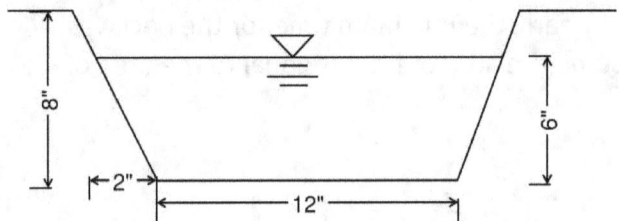

(A) 1.19
(B) 1.67
(C) 5.92
(D) 50.55

Question W18

Three drainage areas have the characteristics as shown below. A retention pond is designed to handle the runoff from these areas. The retention pond has a maximum capacity of 200,000 ft³ and has an average standing volume of 80,000 ft³. Determine the maximum average storm intensity the retention pond can handle without overtopping in 6 hrs. Ignore effects from evaporation and transpiration.

Area 1	Area 2	Area 3
5 acres	4 Acres	6 Acres
Coefficient = 0.16	0.23	0.22

(A) 0.93 in/hr
(B) 1.83 in/hr
(C) 2.01 in/hr
(D) 3.45 in/hr

Question W19

A rectangular concrete open channel is to be designed down a slope to handle a flow rate of water of 500 cfs. The channel is 10' wide and is to be designed for the critical depth of flow. Use a manning's roughness coefficient of 0.015 to determine the required slope of the channel.

(A) 0.002
(B) 0.0046
(C) 0.0055
(D) 0.014

Question W20

A reservoir has an outlet pipe which changes in diameter as shown below. The water at point A has negligible velocity. Assume frictionless flow and water exposed to the air is at atmospheric pressure. Determine the pressure in the discharge pipe at point B if the water exits the pipe at point C.

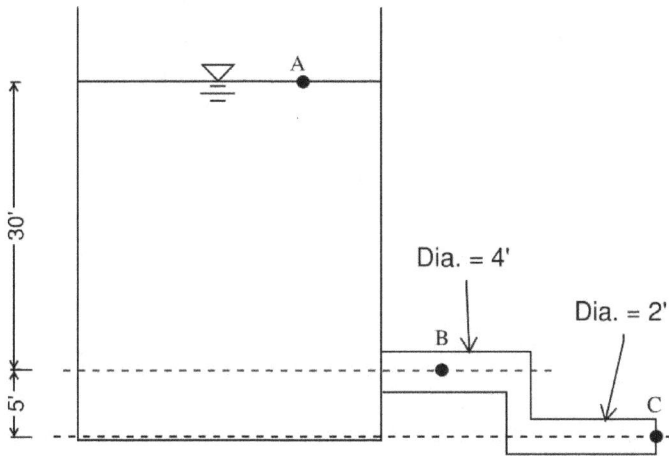

(A) 1734 psf
(B) 1888 psf
(C) 2200 psf
(D) 2862 psf

Question W21

A rainfall event is recorded to have a total runoff of 2 ac-ft and the peak intensity is measured at 1.8 in/hr. The drainage area is farmland with a runoff coefficient of 0.25. if the drainage area is 15 acres total, determine most nearly the unit hydrograph peak discharge.

(A) 1.65 ft³/s-in
(B) 2.22 ft³/s-in
(C) 4.22 ft³/s-in
(D) 6.75 ft³/s-in

Question W22

Two liters of wastewater are analyzed to determine the presence of microorganisms. The concentration of the microorganism is determined to be 9 mg/L and then is measured again 6 hours later to find a concentration of 13.5 mg/L. Determine most nearly the maximum specific growth rate coefficient if the half velocity coefficient is 200 mg/L. The concentration of the nutrients at t = 6 hrs is 5.0 mg/L.

(A) 0.55 hr⁻¹
(B) 1.23 hr⁻¹
(C) 2.01 hr⁻¹
(D) 2.28 hr⁻¹

Question W23

Shown below is a schematic of a network of pipes conveying drainage. The nodes of the system are labeled A through H and the measured flow rates are indicated in cubic feet per second. Given the options below, label the missing flow rates correctly.

| 1.1 |
| 1.2 |
| 3.1 |
| 3.2 |

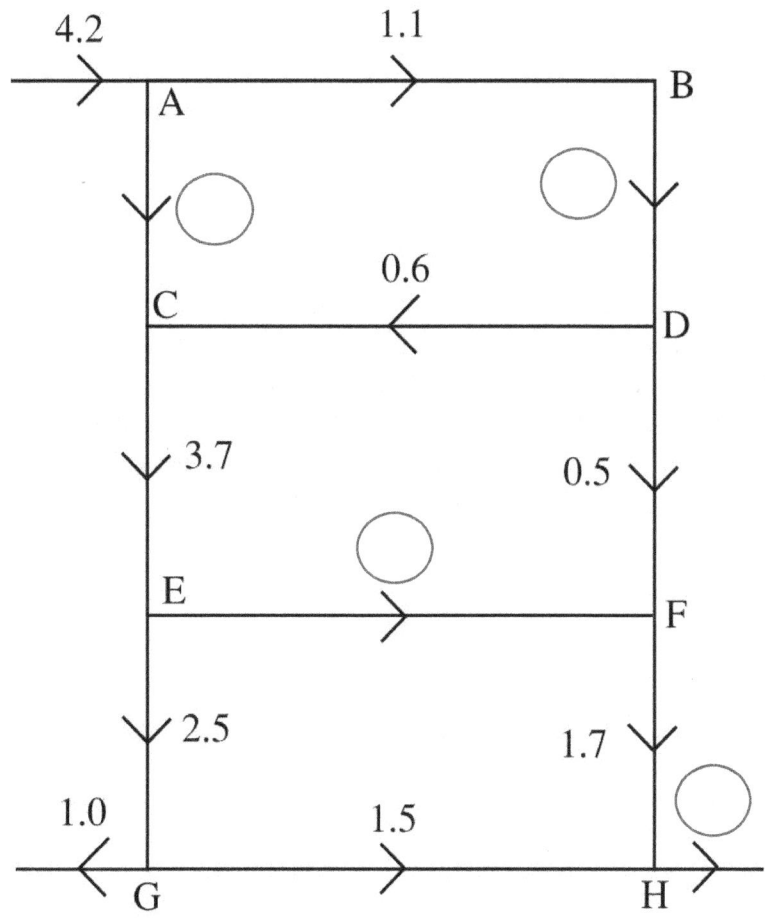

Question W24

A culvert is to be rehabilitated in dry conditions. A cofferdam is placed to stop the flow of water at the inlet and a hydraulic pump at elevation 150' is to be used to divert the water. The water from the pump enters a 6" diameter x 40' long pipe which outlets at elevation 175' and it can be assumed to have a friction factor of 0.02. The calculated flow rate to prevent overtopping of the cofferdam is 4 ft³/s. Determine the horsepower necessary for the pump if it is assumed to run at 80% efficiency.

(A) 20 hp
(B) 50 hp
(C) 100 hp
(D) 250 hp

Question W25

Testing on a river in the Midwest has been performed on the water quality where there has been a decrease in the population of trout. It has been determined that the Dissolved Oxygen content in the river at 60°F is 8.0 mg/L. The river has a flow rate of 5.0 MGD. It is determined that for the trout to sustain their population they should have a DO content of 10.2 mg/L. A water supply will be added to the river to increase the DO content to an acceptable level for the fish flowing at 3.2 cfs. Determine the concentration of dissolved oxygen necessary for the water supply to raise the level in the river for acceptable levels for the fish population.

(A) 8.0 mg/L
(B) 9.9 mg/L
(C) 10.2 mg/L
(D) 15.51 mg/L

Question W26

An existing 8" steel pipe carries drainage water from a catch basin. A differential manometer using mercury is used to record a pressure differential of 1500 lb/ft². To measure the flow of water, a 2" diameter sharp edged orifice is placed in the line. Using a coefficient of contraction of 0.64 and a discharge coefficient of 0.6, determine the flow rate through the pipe.

(A) 0.52 ft³/s
(B) 0.95 ft³/s
(C) 2.56 ft³/s
(D) 10.3 ft³/s

Question W27

A sample of water is determined to have a hardness of 50 mg/L as $CaCO_3$. It is also determined that the sample contains 30 mg/L of HCO_3^- and 20 mg/L of CO_3^{--}. Based on the information provided which of the following statements are appropriate?

(A) All hardness is noncarbonate hardness
(B) All Hardness is carbonate Hardness
(C) Noncarbonate hardness exists
(D) No determination can be made

Question W28

An existing 5' diameter corrugated metal culvert has a friction factor of 0.09 and carries a stream for a length of 50' under a state route. The inlet and outlet are fully submerged with a difference in water elevations of 5'. The velocity is measured to be 6 ft/s. It can be assumed that the pipe entrance has a loss coefficient of 0.5 and the discharge coefficient is 0.98. Determine the discharge of the culvert.

(A) 180 cfs
(B) 220 cfs
(C) 317 cfs
(D) 340 cfs

Question W29

Chlorine gas is added to a sample of water to facilitate disinfection. Which of the following is not a potential result of the reaction if the PH is unknown?

(A) HCl
(B) O_2
(C) HOCl
(D) H^+

Question W30

A rectangular concrete channel is to be designed to handle the flow from an ogee spillway. The spillway has a width of 15' and a height to water of 5'. It is determined that the spillway coefficient is 3.80 $ft^{1/2}$/s. The concrete channel is 20' wide and 8' tall. At the location where the water from the spillway first hits the channel the depth of water is 2' tall. Determine the height of water downstream

(A) 2.0'
(B) 3.1'
(C) 4.1'
(D) 4.7'

Question W31

Which of the following actions would not help to prevent cavitation in pumps?

(A) Increasing the size of pipe
(B) Increase the pump speed
(C) Increasing the size of the pump
(D) Increasing the elevation of the pumped fluid

Question W32

Stream gauge data from a channel is shown below. Five stations are measured having a constant distance between of 4'. Determine the average discharge for this channel from the data.

Station	Depth (ft)	Velocity (ft/s)
1	3	0.3
2	6	0.6
3	10	1.2
4	5	0.8
5	4	0.4

(A) 67.8 cfs
(B) 80.2 cfs
(C) 122.5 cfs
(D) 202.5 cfs

Question W33

Runoff from a watershed is collected to an unpaved swale with a slope of 2%. The watershed has an area of 10 acres and it is calculated that the runoff has a time of concentration before it enters the swale of 30 mins. The required length of the swale to limit the total time of concentration for the watershed and the swale to 35 mins is _____.

Question W34

A rainfall event lasts 12 hrs. Water infiltrates the soil at an initial rate of 10 in/hr. Assuming a decay constant of 2 /hr. Determine the infiltration rate at 25 mins into the storm if the ultimate infiltration rate of 2 in/hr.

(A) 2.35 in/hr
(B) 5.48 in/hr
(C) 6.75 in/hr
(D) 9.22 in/hr

Question W35

Drainage for a road with a concrete gutter (Roughness coefficient = 0.012) is to be designed using a slope of 0.3%. The cross slope of the gutter is 2.5% and the height of water is limited to 3". Determine most nearly the design flow rate for the gutter.

(A) 2.53 cfs
(B) 4.56 cfs
(C) 12.22 cfs
(D) 25.34 cfs

Question W36

A wastewater treatment plant is to be upgraded using new census data from a town. The new population is recorded as approximately 36,000 people. It has been determined that the average daily flow rate of wastewater is 3.6 MGD. The town is to anticipate an increase in flow due to infiltration and inflow of 5%. Determine most nearly the peak flow rate for the design of the system.

(A) 4.4 MGD
(B) 6.3 MGD
(C) 8.5 MGD
(D) 9.1 MGD

Question W37

A retention pond is measured to have an approximate volume of 500,000 gallons. A 10-hr storm event has an average rainfall intensity of 0.4 in/hr and produces a runoff and groundwater inflow combined of 12000 gallons. The expected groundwater outflow is 4250 gallons. The pond is then measured again 12-hrs after the storm and is approximated to have a volume of 521,650 gallons. Ignoring the effects of Transpiration and surface water release, determine the rate of evaporation (in/hr) from the end of the storm until the 2nd measurement if the surface area can be approximated by a 120' diameter circle.

(A) 0.035
(B) 0.067
(C) 0.17
(D) 0.56

Question W38

A volume of water is to undergo some water quality processes. The water has the following characteristics:

Suspended Solids: 220 mg/L
Iron and Manganese concentration: 1.6 mg/L
No noticeable color
No noticeable taste or odor
Chloride 100 mg/L
CaCO$_3$ concentration: 220 mg/L

Select all of the processes that are needed to properly treat the water.

(A) Coagulation and Sedimentation
(B) Rapid Sand Filtration
(C) Salt Water Conversion
(D) Lime Softening
(E) Postchlorination
(F) Aeration

Question W39

A 2050-acre park in an urban area collects runoff and is classified with a curve number of 49. The design storm has a duration of 6 hrs and water travels 4400' on a slope of 2%. Determine most nearly the peak synthetic hydrograph discharge.

(A) 329 cfs
(B) 550 cfs
(C) 1093 cfs
(D) 2290 cfs

Question W40

A wastewater treatment plant is analyzing their budget for the purchase of chemicals on a yearly basis. Chemicals A, B, and C have yearly costs of $10,000, $6,000, and $4,000 respectively. Determine the capitalized cost for the life of the wastewater treatment plant using an interest rate of 8%.

(A) $20,000
(B) $150,000
(C) $250,000
(D) $2,000,000

SOLUTIONS

Solution M1

Use the average end area method

$V = L(A_1+A_2)/2$ where the length between each station is 50'

1+00 to 1+50:

V= 50(0+155)/2=3875 ft³

1+50 to 2+00:

V= 50(155+170)/2=8125 ft³

2+00 to 2+50:

V= 50(170+65)/2=5875 ft³

2+50 to 3+00:

V= 50(65+0)/2= 1625 ft³

Then simply add the volumes and convert to cubic yards

3875+8125+5875+1625 = 19500 ft³ (0.037037 yd³/ft³) = 722.22 yd³. The Answer is **(A)**

Solution M2

A basic knowledge of formwork is needed for this problem. Since the footing has already been poured, formwork will only be needed for all of the faces of the wall except for the top of the wall. The solution is to add up the surface area of the wall faces.

Rear face walls

13.5'(8.0') + 10.0'(8.0') = 188 ft²

Front face walls

(10.0' - 1.5')8.0' + (13.5'-1.5')8.0' = 164 ft²

Outer walls

(1.5'(8.0'))(2 walls) = 24 ft²
Add up the surface areas = 188 + 164 + 24 = 376 ft². The answer is **(C)**

Solution M3

The Critical path of a schedule is the sequence of tasks that determine the minimum amount of time needed to complete the project. Any task on a critical path whose duration is changed will affect the overall schedule of the project.

To solve first develop the task flow chart as shown below. To do this go to each task individually and ask the question "which tasks have this one as a predecessor?". Then draw and connect those tasks with arrows to create a path. Do the same with each task individually until you reach the final task. Then add up the durations for each potential path and the critical path is the largest duration.

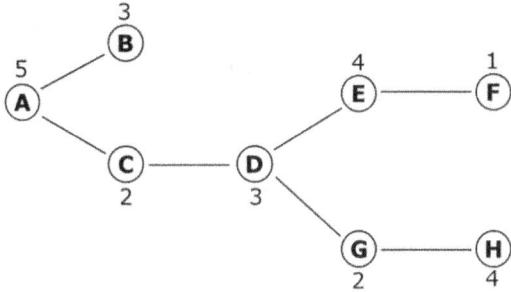

The critical path is tasks A-C-D-G-H with a duration of 16 months.

Solution M4

The initial cost is already at present worth so there is no adjustment of value. The maintenance costs each provide a future value which needs to be converted into a present value

What we have:

n=10 and 20
F= 10,000 and 15,000
i= 3%

What we need: P

Equation needed: $P = F(1+i)^{-n}$

1st maintenance cost: $10,000(1+0.03)^{-10} = 7441$

2nd maintenance cost: $15,000(1+0.03)^{-20} = 8305$ Then add together all present worth costs = 100,000 + 7441 + 8305 = 115746. The answer is **(D)**

Solution M5

First determine the plasticity index: PI = LL − PL = 55-23 = 32

The 50% finer than No. 4 sieve indicates a coarse-grained sandy soil.

Using the word "fines" in a problem is another way to define the No. 200 sieve. Since 20% is greater than 12%, this narrows it down to SM or SC.

A PI over 7 then indicates SC. The Answer is **(A)**

Solution M6

The maximum moment of a simply supported beam is at midspan. Add together the moment from the distributed load and the point load. This can be done by statics or by using the predetermined equations. It's always best to save time by using design aids.

Moment from distributed load = $wL^2/8$ = $2(20^2)/8$ = 100 k-ft
Moment from point load = $PL/4$ = $3(20)/4$ = 15 k-ft
Total moment = 100 + 15 = 115 k-ft. The Answer is **(B)**

Solution M7

The end conditions dictate the effective length of a compression member. From left to right the correct theoretical factors are 2.0, 0.5, 0.7, and 1.0.

Solution M8

The first step is to determine the magnitude and direction of the support reactions

Sum moments about the first support A = 5(5') + 10(10') + 5(15') – B(20') Solve for the right support, B:

B=10 Kips, A=(5 + 10 + 5) – 10 = 10 kips

Then you are able to layout the shear diagram using the following rules:

- To find the magnitude at any point, take a free body diagram from that point to the left-most support and add up the reactions.
- Reactions and loads pointing up are positive, those pointing down are negative
- The shear diagram is flat between concentrated loads
- The shear diagram is sloping along distributed loads

Begin constructing at the left of the beam and work right. The first point is equal to the reaction at A and is upward so the first point is at 10 kips. There is no change until at 5' along the length there is a loading of 5 kips. Therefore, the graph drops to 10 – 5 = 5. Follow this trend to develop the graph.

The answer is **(A)**

Solution M9

First determine the reactions at the left support, A:

Sum moments about point the right support, B = A(10') – 2 k/ft(10')(5') - 5'(2), A = 11 Kips

Since there is no easy equation for the moment at this point on this beam, use the following rule about moment diagrams:

- The magnitude of the moment at any point is equal to the area under the shear diagram curve up to this point

Begin by constructing the shear diagram. We can save time by stopping at point C since we don't need to know any additional information. The graph begins with the reaction at A of 11 kips. You can then find the shear at point C by subtracting the magnitude of the load up to this point.

Shear at C = 11 – 2 k/ft(4') = 3 kips. Therefore, the graph now looks like the figure below. Divide it into sections and find the area:

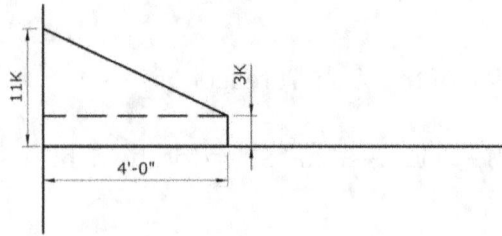

Area 1 = 3 k(4') = 12 kip-ft
Area 2 = ½ (11k - 3k)(4') = 16 kip-ft   Total Moment = 12 + 16 = 28 K-ft   The answer is **(C)**

Solution M10

6" diameter concrete cylinders are a common method of testing concrete compressive strengths. The stress from an axial load is determined by P/A.

Area of each cylinder = πr^2 = $\pi(3)^2$ = 28.27 in^2

Cylinder 1 stress = 102000/28.27 = 3607 psi
Cylinder 2 stress = 111000/28.27 = 3925 psi
Cylinder 3 stress = 100500/28.27 = 3554 psi

Average of the cylinders = (3607 + 3925 + 3554)/3 = 3695 which is most nearly 3700 psi. The answer is **(D)**

Solution M11

You can find the length of the curve by subtracting stations L = 2+25 – 1+50 = 75'
Convert the interior angle to a decimal 15'/60 = 0.25 Therefore I = 10.25°
The radius can be found by L = ($2\pi rI$)/360° = 75' = $2\pi r$(10.25)/360 solve for r = 419.2'

The answer is **(B)**

Solution M12

The clearance of the bridge is simply the elevation difference between the low point of the bridge and the elevation of the roadway at that station.

First calculate the gradient of the curve A = (G_2-G_1)/L = (-4-2)/8 = -0.75

Determine the beginning of vertical curve (BVC) station:

BVC = 10+00 – 800/2 = 6+00

Determine the elevation of BVC = 155.0 – (.02)400 = 147.0'

Determine the elevation of station 8+00 by using the equation of a parabola (Be careful with signs) where X is the difference in stations from the point of interest to the BVC

$Y = Elev_{BVC} + G_1(X) + (A/2)X^2 = 147.0 + 2(2) + (-0.75/2)(2)^2 = 149.5'$

Then subtract the road elevation from the low point of the bridge to find the clearance

170.0 – 149.5 = 20.5'. The answer is **(B)**

Solution M13

The low point is where the slope of the gradient is zero. Therefore:

$X = G_1/A$

$A = (G_2-G_1)/L = (2-(-3))/6 = 0.833$

$X = 3/0.833 = 3.6$ stations

The low point is at station 7+00 + 3+60 = 10+60.

Solution M14

Use the equation for the middle ordinate of a horizontal cure

$M = R(1-\cos(28.65S/R)) = 700(1-\cos(28.65(220)/700)) = 8.63'$

The answer is **(A)**

Solution M15

Working backwards from compaction, apply adjustment factors at each stage and carry over the volume:

To achieve 100 yards of 90% compaction you will need $100/0.9 = 111.11$ yd^3

5% is lost so you will need an additional 5% to account for this $111.11(1.05) = 116.65$ yd^3

A swell factor is the volume of the loose excavation material to the in-place excavation material. To find the in-place material, divide by the swell factor $116.65/1.07 = 109.03$ yd^3

The answer is **(D)**

Solution M16

To solve mix design problems, follow the units.

Determine the weight needed in water $W_{water} = W_{cement}(w/c \text{ ratio}) = 500(0.65) = 325$ lb/yd^3

The specific weight of water can always be taken as 62.4 lb/ft^3. Use the specific weight to convert a weight to a volume

$V_{water} = W_{water}/\gamma_{water} = (325 \text{ lb/yd}^3)/(62.4 \text{ lb/ft}^3) = 5.21$ ft^3/yd^3 The answer is **(A)**

Solution M17

There are 3 conditions which will simplify the needed equations.

1. Friction is neglected between the wall and soil. This make the angle of external friction ∂=0
2. The backfill is horizontal. This makes the slope of backfill β=0
3. The wall face is vertical. This makes θ=0

The equations needed are now

The active earth pressure coefficient $k_a = \tan^2(45° - \phi/2)$
The total active resultant $R_a = \frac{1}{2} k_a \gamma H^2$

$k_a = \tan^2(45 - 25/2) = 0.406$
$R_a = \frac{1}{2}(0.406)(115 pcf)(15.5)^2 = 5606$ lbs

The force is applied at H/3 = 15.5/3 = 5.167'

The moment is 5.606k(5.167') = 28.9 k-ft The answer is **(C)**

Solution M18

The void ratio is the volume of the voids divided by the volume of the solids.

If a sample is saturated it can be assumed that there are no air voids and thus the total volume of the voids is equal to the volume of the water in that state.

When a sample is dried, there is no more water left in the sample and we have simply the weight of the soil.

Therefore, we are left with the following values

W_W = 50 – 42 = 8 lbs

W_S = 42 lbs

Since the density of water is known we can then convert the weight to volume

V_W = 8/62.4 = 0.128

And since the sample was saturated, all voids are filled with water and $V_W = V_V$

We then need to determine the volume of soil from the weight. Since we have the Specific Gravity (SG), we can determine the density by multiplying by the density of water

$\gamma = 2.4(62.4) = 150$ lb/ft³

$V_s = 42/150 = 0.28$

Void Ratio = $0.128/0.28 = 0.46$ The answer is **(D)**

Solution M19

The total distance is a sum of two components. The first is before breaking and the second is after breaking. This is represented in the following equation:

$S_{stopping} = vt_p + S_b$

The first component assumes the velocity is constant during perception reaction and is simply (100 ft/s)(2 seconds) = 200 ft

The stopping distance when breaking occurs is the following:

$S_b = v^2_{mph}/(30(f+G))$

Since the slope is on an incline G is positive. If it was on a decline, it would be negative.

Convert the velocity from ft/s to mi/hour

100ft/sec(3600 sec/hr)/5280 ft/mi = 68.2 mi/hr

$S_b = (68.2^2)/(30(0.30+0.02)) = 484.5$ ft

$S_{stopping} = 200 + 484.5 = 684.5$ ft The answer is **(B)**

Solution M20

A 100-yr storm event is a storm of a design intensity which is only anticipated to occur once every 100 years.

The answer is **(C)**

Solution M21

Use the Darcy equation:

$h_f = (fLv^2)/(2Dg)$

Use 32.2 ft/sec² as g and then plug in and solve for v

$3.0 = (0.02)(500)v^2/2(2)(32.2)$

v = 6.2 ft/s. The answer is **(C)**

Solution M22

The appropriate equation in open channel flow is the Chezy-Manning equation:

$Q = (1.49/n)AR^{2/3}S^{1/2}$

First calculate the hydraulic radius which is the area of water divided by the wetted perimeter which is the perimeter of the sides of the channel which are in contact with water.

R = (6*3)/(6+3+3) = 1.5

$Q = (1.49/0.015)(6*3)(1.5)^{2/3}0.003^{1/2}$ = 129.3 The answer is **(A)**

Solution M23

For the given information, the Hazen-Williams equation is appropriate:

$h_f = 10.44LQ^{1.85}/C^{1.85}d^{4.87}$

$40 = 10.44(100)Q^{1.85}/(140)^{1.85}(3)^{4.87}$

Solve for Q = 432 gpm The answer is **(D)**

Solution M24

The time base is the amount of time that the flow exceeds the base flow. The answer is **(C)**

Solution M25

The flow rate of the pipe can be used to determine the area required to limit the flow velocity

The flow rate of the pipe can be determined by the conservation of flow principle

$Q_1 + Q_2 = Q_3$

The flow of the drainage areas can then be determined by the Rational method:

Q = ACi Therefore Q_3 = 15(0.18)(1.5) + (10)(0.22)(1.5) = 7.35 cfs

Use the calculated velocity to find the required area Q/V = A = 7.35/0.26 = 28.28 ft²

The answer is **(A)**

Solution M26

When determining how many 0-Force members a truss has, analyze each joint individually as a free body diagram and follow these guidelines:

1. In a joint with 2 members and no external forces or supports, both members are 0-force
2. In a joint with 2 members and external forces, If the force is parallel to one member and perpendicular to the other, then the member perpendicular to the force is a 0-force member.
3. In a joint with 3 members and no external forces, if 2 members are parallel then the other is a 0-force member

Therefore, each joint can be analyzed as follows:

Joint A and H – These joints have 2 members and an external force so refer to guideline 2. Neither member parallel to the force therefore all members are non-zero

Joint C, D, and G – These joints have 3 members and no external force so refer to Guideline 3. 2 members are parallel therefore the other is a 0-Force member. Therefore, members C-B, D-E, and F-G are 0-Force members

Joints B, E, and F – These joints do not meet any of the guidelines. Therefore, all members are non-zero

Solution M27

This question is provided simply to illustrate that the analysis of the external forces of a truss can be determined in the same way as beams. There may be a tendency to perform unnecessary computations. Simply apply basic statics:

Sum the Forces about H = 0 = -40A + 5(30) + 5(10), Solve for A = 5 Kips

The Answer is **(B)**

Solution M28

First determine the reaction at A. See solution M27.

Determine the angle of triangle A-B-C $Tan^{-1}(8/10) = 38.66$

Using the method of joints, analyze joint A using a free body diagram

Sum the forces in the Y-Direction 5 = ABsin(38.66), AB = 8.00

Sum the forces in the X-Direction AC = ABcos(38.66) = 8cos(38.66) = 6.24

Analyze Joint C

Sum the Forces in the X-Direction AC = CE = 6.24

The answer is **(C)**

Solution M29

Rebar corrosion is most often the result of a chemical reaction due to the intrusion of chlorides into existing concrete. The answer is **(A)**

Solution M30

The water to cement ratio is inversely proportional to the concrete strength and therefore has a direct impact. The answer is **(A)**

Solution M31

Use the Bernoulli equation for the conservation of energy

$E_t = E_{pr} + E_v + E_p = p + v^2/2g + z$

At the water surface of the reservoir, the total energy is the potential energy only and $E_t = 0 + 0 + 50 = 50$

Due to the conservation of energy the total energy at the water surface is equal to the total energy at the exit of the pipe. Therefore, the total energy at the exit
$E_t = 50 = 0 + v^2/2(32.2) + 0$, solve for v = 56.75 ft/s The answer is **(B)**

Solution M32

The answer is **(D)**.

The other 3 assumptions are essential to the Bernoulli equation.

Solution M33

Use the conservation of flow principle. $Q_1 + Q_2 = Q_3$ and therefore $A_1V_1 + A_2V_2 = Q_3$

$Q_3 = 3.0(1.2) + (4.0)0.8 = 6.8$ ft³/s The answer is **(D)**

Solution M34

The effective stress is the density times the height of each level. However, if the water table is present, the density is reduced by that of water:

Effective Stress = 130(20) + 100(10) + (100-62.4)(10) = 2600 + 1000 + 376 = 3976 psf

The answer is **(C)**

Solution M35

The Maximum bending stress is determined at the location of maximum moment.

First determine the reactions at the supports

Sum moments about B = 0 = -15A + 10(10) + 10(5), A = 10

Due to symmetry A = B = 10

The Max moment for a beam with 2 symmetrical point loads = PX where X is the distance to the first point load = 5'(10 kips) = 50 Kips-ft

Then apply the equation for bending stress Mc/I

Moment of inertia of a rectangle $(1/12)bh^3 = (1/12)(4)(6)^3 = 72$ in^4

The bending stress = (50 k-ft)(12 in/ft)(3 in)/72 in^4 = 25 ksi

The answer is **(A)**

Solution M36

P/A = 0.1 ksi = 9/(12w), w = 7.5"
B can be determined by finding the additional horizontal distance from the critical section for shear to the base.

$$\frac{1}{8} = \frac{x}{6} \quad x = 0.75"$$

b = 7.5" + 0.75" = 8.25"

The answer is **(C)**

Solution M37

The determination of which roadside safety barrier system is determined by the distance from the edge of road. Barriers can be flexible, semi rigid, or rigid. A concrete barrier is rigid and is for locations where zero deflection is needed. A metal beam rail is semi-rigid and is used for objects or drop-offs within a short distance of the road. Cable systems are flexible and have the most deflection and are used for objects or drop-offs further from the road.

The answer is **(C)**

Solution M38

The peak hourly traffic volume is the hour-long timeframe in which the most cars are observed. First calculate the total volume for each hour-long time frame:

8:00-9:00 = 500 + 560 + 650 + 625 = 2335
8:15-9:15 = 560 + 650 + 625 + 630 = 2465
8:30-9:30 = 650 + 625 + 630 + 600 = 2505
8:45-9:45 = 625 + 630 + 600 + 540 = 2395
9:00-10:00 = 630 + 600 + 540 + 460 = 2230

Then determine the largest volume = 2505 The answer is **(C)**

Solution S39

The equation for deflection at the end of a cantilever beam with a point load is:

$Pl^3/3EI$

First determine the moment of inertia:

$I = (1/12)(6)12^3 = 864 \text{ in}^3$

$\Delta = 1(10(12))^3/(3(3605)(864)) = 0.18"$, The answer is **(B)**

Solution M40

The specific gravity is the density of the soil over the density of water.

The volume of the voids can be determined by subtracting the soil volume from the total

$V_V = V_T - V_S = 2.0 - 1.5 = 0.5 \text{ ft}^3$

The degree of saturation can be used to determine the volume of water

$V_W = V_V(S) = 0.5(0.75) = 0.375 \text{ ft}^3$ Convert volume to weight $W_W = 0.375(62.4) = 23.4 \text{ lb}$

Then determine the weight of soil thought the moisture content

$W_S = W_W/w = 23.4/0.1 = 234$ lbs, the density is then $234/1.5 = 156$

The SG = 156/62.4 = 2.5 The answer is **(B)**

Solution M41

First, the % passing of fines is another term for the % passing the no. 200 sieve. Using the AASHTO Classification chart, this narrows the category down to the A-2 class. Then determine the Plasticity Index:

Plasticity Index = Liquid Limit – Plastic Limit = 36 – 24 = 12

Then using the LL=36 and PI=12, the appropriate Class is A-2-6.

The Answer is **(C)**

Solution M42

The Charpy V-Notch is a test for the toughness of steel.

The answer is **(D)**

Solution M43

First categorize the loads as dead or live loads:

Dead: Floor Slab, Flooring, Utilities. These are loads which are permanent to the structure.

Live: Pedestrian, Furniture. These are loads which will or may move.

Then calculate the load:

1.2(100 + 10 + 2) + 1.6(80 + 40) = 326.4 psf

Solution M44

This beam is fixed on one end and supported on the other with a uniform distributed load. Use the beam chart to find the appropriate equation:

$$\Delta = \frac{wL^4}{185EI}$$

First find the moment of inertia:

I = 1/12bh³ = 1/12(6)(12)³ = 864 in⁴

Then plug into the equation:

$$\Delta = \frac{(2)(10*12)^4}{185(3605)(864)} = 0.72"$$

The answer is **(D)**

Solution M45

The work which takes place within the adjacent property is temporary in nature and the current owner will maintain rights of the property and therefore a temporary easement would be required.

The answer is **(A)**

Solution M46

Determine the quantities for each item in the appropriate unit.

Concrete Wall = (10(1) + 8(2.5))20 = 600 cu. ft.

Backfill = ((4)10 + ½(2)(10))20 = 1000 cu. ft./27 = 37.04 cu. yd.

Drain = 20'

Then use the unit prices to find the total cost:

600(30) + 37.04(100) + 20(20) = 22104.

The answer is **(B)**

Solution M47

The answer is **(B)**

Solution M48

First convert the load in tons to lbs. 75 tons (2000 lbs/ton) = 150,000 lbs

Then using the chart for a 120' boom, the most appropriate answer is 30'.

The answer is **(B)**

Solution M49

Due to the desired accelerated time of construction, cast in place can be eliminated due to the need for curing times. Then the best choice is dictated by the span length. 80' is much longer than the capacity for precast concrete and post tensioned concrete is not economical at this span. Therefore, the best choice is Prestressed-precast concrete.

The answer is **(C)**

Solution M50

The length L can be determined by setting the reaction at the temporary support to the maximum allowable of 5 kips. Sum moments about the end of the beam:

0 = 0.288 k/ft(30')(30'/2) – L(5 K) Solve for L = 25.92'

The answer is **(C)**

Solution M51

The need for Temporary earth retaining is based on the soil type and the depth of cut. The depth of cut without TERS is limited to less than 5 feet. Also stable rock is a type of soil that is considered sufficiently safe without support. Therefore the answer is A

The answer is **(A)**

Solution M52

$$F = \frac{N_o c}{\gamma_{eff} H}$$

$$\gamma_{eff} = 125 - 62.4 = 62.6 \, pcf$$

$$1.5 = \frac{6.2(300)}{(62.6)H} \quad H = 19.8'$$

X = 19.8/1.5 = 13.2

Available Right of way = 100 − 2(13.2) = 73.6′

The answer is **(C)**

Solution M53

$$h = \Sigma k \frac{v^2}{2g}$$

$$K_{90} = 0.9$$

$$K_c = 0.5\left[1 - \left(\frac{D_1}{D_2}\right)^2\right] = 0.5\left[1 - \left(\frac{8}{12}\right)^2\right] = 0.277$$

$$A = \pi r^2 = \pi 0.33^2 = 0.4 \, ft^2$$
V = Q/A = 2.0/0.4 = 5 ft/s

$$h = (0.9 + 0.277)\frac{5^2}{2(32.2)} = 0.46'$$

The answer is **(B)**

Solution M54

The Euler critical stress is the following equation:

$$F_{Cr} = \frac{\pi^2 E}{\left(\frac{KL}{r}\right)^2}$$

Where r = Radius of Gyration = $\sqrt{\frac{I}{A}}$

I = 1/12(36)(36)³ = 140000 in⁴

$$r = \sqrt{\frac{140000}{36(36)}} = 10.4 \ in$$

The K can be determined by the end conditions, for fixed and pinned the design value is 0.8 from the chart.

$$F_{Cr} = \frac{\pi^2(3605)}{\left(\frac{0.8((25)(12))}{10.4}\right)^2} = 66.8 \ ksi$$

The answer is **(D)**

Solution M55

Primary consolidation is a gradual consolidation as water leaves the voids over time. This is mostly an issue in clay type soils. Therefore, CH and CL can be expected to see the most primary consolidation.

Solution M56

The superelevation transition distance is the distance between the beginning of the transition to superelevation to the point of being fully superelevated which is the combination of the tangent (also known as crown) runout and the superelevation runoff. The correct labels are:

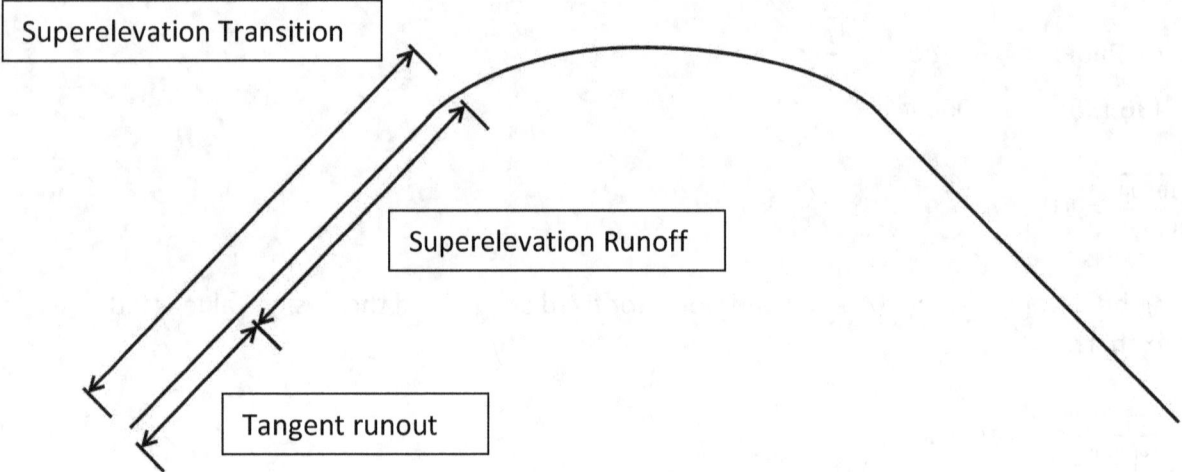

Solution M57

The flow in needs to equal the flow out of a system. However since the flow out is limited, this will produce a backup of water which will begin to fill the pond. First determine the flow rate of the outlet:

$Q = VA = (0.25)(\pi r^2) = (0.25)(\pi(1)^2) = 0.785$ ft³/s

Then the excess can be determined by subtracting the flow out from the flow in = 0.85 – 0.785 = 0.065 ft³/s

Then determine the size of the pond by determining how much volume of water will accumulate during the storm:

V = 0.065(6 hr)(60 min/hr)(60 s/min) = 1404 cu. ft. = 52 cu. yd.

The answer is **(B)**

Solution M58

The N-value is the number of blows required to drive the sampler 12" after the initial 6". Therefore, the numbers to average are 26, 38, and 60. The average is 41.33.

The answer is **(D)**

Solution M59

The permeability of soil is determined by Darcy's Law:

Q = KiA

First the flow rate can be determined from the sample by dividing the volume by the time:

Q = 1/10 = 0.1 in³/s

Then determine the cross sectional area of the sample:

A = πr^2 = $\pi(3)^2$ = 28.27 in²

The hydraulic gradient, I, is determined by the difference in pressure head divided by the length;

i = 3/12 = 0.25

Then solve for K = Q/iA = 0.1/((0.25)(28.27)) = 0.014 in/s

The answer is **(B)**

Solution M60

For different aspects of construction, the required level of accuracy varies. For the construction of bridges this is 0.01. Therefore, the measurement should be recorded as 158.57'.

The answer is **(B)**

Solution M61

When a project is complete, areas which have been excavated should be reestablished with seeding to promote growth. The other methods are temporary measures during construction.

The answer is **(C)**

Solution M62

The compressive stress for a concentric point load is P/A. However since this is eccentric, an additional stress due to the moment must be applied. Therefore:

$$f_c = \frac{P}{A} + \frac{MC}{I}$$

I = 1/12(14)(14)³ = 3201 in⁴

$$f_c = \frac{5}{14(14)} + \frac{5(6)(7)}{3201} = 0.091 \; ksi$$

The answer is **(A)**

Solution M63

The choice of an appropriate unit of measure needs to be a representation of what varies from project to project as well as what can be easily measured. In this case the box culverts have a consistent area and therefore a measure of area would not represent the work appropriately. The unit each would be difficult to estimate since box culvert sections can vary based on the manufacturers capabilities. This would be difficult to approximate the quantity. Lump sum is used for general tasks which may be difficult to measure. What does vary, is known, and can be easily measured is the length of the box culverts. Therefore the appropriate unit is Linear ft.

The answer is **(A)**

Solution M64

The critical section for moment on a spread footing is at the face of the column. For a point load only the equation for the moment due to the uniform bearing pressure is:

$M_u = q_u L l^2/2 = (50/(10(10)))(10)(4)^2/2 = 40$ K-ft   The answer is **(D)**

Solution M65

The time of concentration is a product of three components. However in this example we are only looking for sheet flow time. Therefore:

$$t_c = t_{sheet} = \frac{0.007(nL_o)^{0.8}}{\sqrt{P_2} S_{deciaml}^{0.4}} = \frac{0.007((0.15)(250))^{0.8}}{\sqrt{2.2}(0.05)^{0.4}} = 0.284 \; hrs$$

The answer is **(A)**

Solution M66

The Peak Hour Factor is ratio of the hour interval with the greatest volume to four times the largest 15 min. interval volume. First determine the hour interval with the greatest volume

7:00-8:00	1900+2200+2350+2000=8450
7:15-8:15	2200+2350+2000+1700=8250
7:30-8:30	2350+2000+1700+1200=7250

Therefore the largest hourly volume is 8450 in interval 7:00-8:00. The 15-min. interval with the greatest volume in this interval is 7:30-7:45 with a volume of 2350.

$$PHF = \frac{V_{vph}}{4V_{15\,min,peak}} = \frac{8450}{4(2350)} = 0.8989$$

The answer is **(C)**

Solution M67

Use the NRCS method

S = 1000/CN − 10 = 1000/79 − 10 = 2.658

$$Q = \frac{(3.5 - 0.2(2.658))^2}{3.5 + 0.8(2.658)} = 1.56\ in$$

The answer is **(A)**

Solution M68

Use the Terzaghi Bearing Capacity equation:

$$q_{ult} = \frac{1}{2}\gamma B N_\gamma S_\gamma + cN_c S_c + (p_q + \gamma D_f)N_q$$

However, since the soil is cohesion less and there is no surcharge load, the equation becomes as below. Also since the footing is a strip type, the shape factors are 1.0 and the equation becomes:

$$q_{ult} = \frac{1}{2}\gamma B N_\gamma + (\gamma D_f)N_q = \frac{1}{2}(0.125)(8)(2.5) + (0.125)(2.5)(4.4) = 2.625\ ksf$$

The ultimate pressure then needs to be adjusted for overburden:

$q_{net} = q_{ult} - \gamma D_f = 2.625 - 0.125(2.5) = 2.3125$ ksf

Finally calculate the allowable from the net pressure:

$q_a = q_{net}/FS = 2.3125/2.5 = 0.925$ ksf

The answer is **(A)**

Solution M69

A mat foundation is a large scale pour which would require a slower curing process to limit cracking and reduce heat release. The types of cement which are appropriate for this are type II and IV. Type II is more suitable for protection against sulfate attack and therefore the choice is Type IV.

The answer is **(D)**

Solution M70

V = Area under Hydrograph

Lag Time = Time from greatest rainfall to peak discharge = Time from hour 2 to hour 6

1 hr (60)(60) = 3600 seconds

3600(1/2(0.5) +1.4 +2.5 + 6.6 + ½(8.6)) = 5480 cf

The answer is **(B)**

Solution M71

To prevent cracking, the modulus of rupture needs to be greater than the applied stress. First determine the maximum moment in the wall panel. Since it can be approximated as a simply supported beam with uniform load, the max moment is:

$$M_u = \frac{wl^2}{8} = \frac{(0.8)(18)^2}{8} = 32.4k-ft = 388.8k-in$$

Calculate the moment of inertia of the resisting section:

I = 1/12((8)12)(8)³ = 4096 in⁴

Then determine the stress due to bending:

$$f_b = \frac{Mc}{I} = \frac{(388.8)(4)}{4096} = 0.380 \; ksi = 380 \; psi$$

Then equate the stress to the equation for the modulus of rupture to find the minimum strength:

$7.5\sqrt{f'_c} = 380$

$f'_c = 2567$ psi

The answer is **(C)**

Solution M72

In this example the present worth of each project is known and the future value at different times is needed. The equation to turn a present value to a future value is:

$F = P(1+i)^n$

Use this equation for each project:

$F_1 = 100000(1 + 0.03)^3 = 109273$

$F_2 = 175000(1 + 0.03)^4 = 196964$

$F_3 = 225000(1 + 0.03)^5 = 260837$

Then add up the total cost = 567073.

The answer is **(D)**

Solution M73

For flexure the important factor is the moment of inertia. Unlike the other members, the WT shape does not have a bottom flange portion and has a poor distribution of area. Therefore this would be the least effective in flexure.

The answer is **(B)**

Solution M74

The strength of the steel has no effect on the steels ability to resist corrosion. Therefore High Strength Steel does not assist in corrosion protection.

The answer is **(C)**

Solution M75

The sandy soil creates a stable bearing condition which indicates that a deep foundation element is not necessary. Also note that the bridge does not cross a waterway. Often for scour prevention measure, bridges which cross water are required to be on deep foundation elements. Therefore, a spread footing should be sufficient.

The answer is **(A)**

Solution M76

Because of limited resources, the crews have to be assigned work. Crews can work independently or simultaneously.

A = 4 days per crew/2 crews = 2 days

B + C = 12/2 = 6 days

D + E = 18/2 = 9 days

F = 10/2 = 5 days

Total = 2 + 6 + 9 + 5 = 22 days

Total cost of labor = 22(8 hrs/day)(50 $/hr)(4 workers)(2 crews) = $70400

The answer is **(C)**

Solution M77

For the existing schedule the breakdown of paths and durations are as follows:

Path	Total Duration
A-B-C-E	22
A-B-C-F	21
A-B-D-F	19
A-B-D-G	21

Therefore the critical path is A-B-C-E. The durations will then change and the new critical path needs to be determined. The revised schedule is shown below:

Path	Total Duration
A-B-C-E	23
A-B-C-F	25
A-B-D-F	26
A-B-D-G	25

The critical path now becomes A-B-D-F with a duration of 26.

Solution M78

In the design of slabs, the analysis is performed by taking a 1' wide strip of the slab as a beam. Therefore, this can be analyzed by taking a 1' wide beam which is fixed at both ends with a distributed load. The maximum moment for this loading and support condition is:

$$M = \frac{wl^2}{12} = \frac{(0.5\ ksf)(1')(20)^2}{12} = 16.67\ k-ft$$

The answer is **(D)**

Solution M79

Offset distance to the Baseline of road is indicated as 16'. This point has an elevation of 144'

Distance from centerline of road to face of curb = 16' – 4' =12'

Calculate drop in elevation from the cross slope 144' – 0.02(12) = 143.76'

Add height of curb 143.76 + 10/12 = 144.60'

The answer is **(C)**

Solution M80

Find the yield strength by identifying the point at which the load decreases and there is an appreciable change in elongation compared to the increase in load in this case it can be identified at load 2656 lbs.

Strength = $P/A = 2656/(\pi r^2) = 2656/(\pi(0.125)^2) = 54.1$ ksi

The answer is **(C)**

Solution W1

To solve this, you must find your way to the calculated basin depth which is the depth of the water from the given characteristics. First convert the flow from MGD to ft³/s:

$$\left(\frac{1.2 \times 10^6 \, gal}{day}\right)\left(\frac{day}{24 \, hr}\right)\left(\frac{hr}{3600 \, s}\right)\left(\frac{ft^3}{7.48 \, gal}\right) = 1.86 \, cfs$$

Convert the detention time to seconds:

$$2 \, hrs \left(\frac{60 \, mins}{hr}\right)\left(\frac{60 \, s}{min}\right) = 7200 \, s$$

The detention time can then be used to determine the volume of water in the clarifier and therefore the depth of water:

$$t_d = \frac{V}{Q} = 7200 = \frac{45(35)(D)}{1.86} ; D = 8.5'$$

Then add 6' to get the minimum height: H = 8.5 + 6.0' = 14.5'

The answer is **(B)**

Solution W2

First using the given parameters, the influent BOD concentration, S_o, can be determined:

$$F:M = \frac{S_{o,mg/L} Q_{o,MGD}}{V_{o,MG} X_{mg/L}} = 0.2 = \frac{S_o(5.0)}{(1.57)(1500)}; \, S_o = 94.2 \, mg/L$$

Then use the equation for removal efficiency to determine the effluent:

$$\eta = \frac{S_o - S}{S_o} = 0.9 = \frac{94.2 - S}{94.2}; S = 9.42 \, mg/L$$

The answer is **(A)**

Solution W3

Risk from a substance is calculated by the following equation:

$$Risk = \frac{(Concentration)(Intake)(Absorption\ Factor)(Exposure)(Risk\ Factor)}{(Body\ Weight)(Lifetime)}$$

Therefore you can plug in the information as appropriate to find the concentration:

$$0.003 = \frac{C\left(1.2\frac{L}{D}\right)(0.95)(10\ years \times 365\frac{days}{year})(0.4\ kg * \frac{day}{mg})}{(68\ kg)(75\ years \times 365\frac{days}{year})}; C = 3.36\ mg/L$$

The answer is **(C)**

Solution W4

There are two necessary concepts here related to parallel pipes. First the head loss in a parallel pipe network is equal among all pipes which are parallel. And second the flow rate is the sum of the flow rate from all pipes. Therefore, the head loss between pipes 1 and 2 can be used to find the velocities. Use the Hazen Williams equation to determine the head loss in pipe one:

$$h_f = \frac{3.022\ ^{1.85}L}{C^{1.85}D^{1.165}} = \frac{3.022(6.0)^{1.85}(45)}{(80)^{1.85}(2)^{1.165}} = 0.5'$$

From the principles of parallel pipes we know that $h_{f1} = h_{f2} = 0.5'$

We can rearrange the Hazen Williams equation to find the velocity in pipe 2 as follows:

$$v = \frac{0.55CD^{0.63}h_f^{0.54}}{L^{0.54}} = \frac{0.55(90)3^{0.63}0.5^{0.54}}{55^{0.54}} = 7.81\ ft/s$$

Then the total flow for the system is the sum of the two pipes $Q = V_1A_1 + V_2A_2 = 6.0\pi 1^2 + 7.81\pi 1.5^2 = 74.05$ cfs

The answer is **(A)**

Solution W5

In this problem we need to determine the flow rate for the supplemental pipe to determine the area. Since it will be 50% of the existing pipe, we must use the pitot gauge to determine the velocity of the existing flow which we can then determine the flow rate. The equation for velocity in a pitot static gauge is:

$$v = \sqrt{\frac{2gh(\rho_m - \rho)}{\rho}} = \sqrt{\frac{2(32.2)(\frac{2}{12})(848.6 - 62.4)}{62.4}} = 11.63 \text{ ft/s}$$

Then determine the flow rate of the existing pipe $Q = VA = 11.63\pi(1)^2 = 36.54 \text{ ft}^3/\text{s}$

The supplemental pipe will take 50% of this flow $Q_{supp} = 36.54(0.5) = 18.27 \text{ ft}^3/\text{s}$

$A_{supp} = Q/V = 18.27/11.63 = 1.57 \text{ ft}^2$

The answer is **(A)**

Solution W6

First find the average DO Content for both the initial and incubated samples by finding the arithmetic mean:

$DO_f = (2.1 + 2.3 + 2.4)/3 = 2.27$ mg/L; $DO_i = (6.2 + 6.8 + 6.4)/3 = 6.47$ mg/L

The you can determine the biochemical oxygen demand at the 5 day time period:

$$BOD_5 = \frac{DO_i - DO_f}{\frac{V_{Sample}}{V_{Sample} + V_{Dilution}}} = \frac{6.47 - 2.27}{\frac{4}{4 + 250}} = 267 \text{ mg/L}$$

Then you can determine the Ultimate BOD:

$$BOD_5 = BOD_u(1 - 10^{-K_d t}) = 267 = BOD_U(1 - 10^{(-0.11)(5)}); \quad BOD_U = 371.8 \text{ mg/L}$$

The answer is **(C)**

Solution W7

First convert pounds of Aluminum Sulfate to grams:

5 lbs(453.6 g/lb) = 2268 g

Then convert to moles using the molar mass:

2268 g(1g/342.15 mols) = 6.63 mols

Then use the chemical equation for Phosphorous removal to determine the coefficient ratio for the balanced equation:

$$Al_2(SO_4)_3 + 2PO_4 \rightarrow 2AlPO_4 + 3SO_4$$

Therefore 2 mols of PO4 are needed for every mol of Aluminum Sulfate.

Moles of PO4 = 2(6.63) = 13.25 mol; Convert to grams = 13.25(94.97) = 1259.0 g

Then determine the mass of phosphorous by the percentage: 1259(.3262) = 410.7 g

And convert to pounds: 410.7 g/(453.6 g/lb) = 0.91 lbs

The answer is **(A)**

Solution W8

Use the equation for appropriate doses of a coagulate:

$$F_{lb/day} = \frac{D_{mg/L} Q_{MGD} \left(8.345 \frac{lb-L}{mg-MG}\right)}{PG}$$

Remember the availability can be taken as 1.0 if there are no supply issues:

$$50 = \frac{D_{mg/L} 2.1 \left(8.345 \frac{lb-L}{mg-MG}\right)}{(0.45)(1.0)}$$

Solve for D = 1.28 mg/L

The answer is **(A)**

Solution W9

Cross sections have ideal dimensions to most efficiently carry water. For rectangles, this is achieved by setting the depth equal to the width/2. Therefore, the third option with a width of 8 and a depth of 4 is the correct choice.

Then, for trapezoids, the ideal section is when the depth is equal to two times the Hydraulic Radius. If we take option C and calculate:

Depth = 4cos(30) = 3.5'

R = Area/Wetted Perimeter

$$A = \frac{a+b}{2}h = \frac{4 + (4 + 4\cos(60)\,2)}{2}(3.5) = 21$$

R = 21/(4+4+4) = 1.75

2R = 1.75(2) = 3.5 = d

Therefore, the first option is an efficient section

Solution W10

The equation for the mean cell residence time is as follows:

$$\theta = \frac{VX}{Q_e X_e + Q_w X_w}$$

Because of the conservation of mass, the flow in must equal the flow out of the tank. Therefore, we must divide the flow out by the indicated percentages and the equation becomes:

$$10 = \frac{V(2000\frac{mg}{L})}{0.85(1062\ cfd)(30\frac{mg}{L}) + 0.15(1062\ cfd)(2200\frac{mg}{L})} ; V = 1887\ ft^3$$

Then use the area to find the depth:

V = dπr² = 1887 = dπ(10)² ; d = 6.0'

The answer is **(A)**

Solution W11

First given the flow rate you can determine the velocity of flow at point A:

$V_A = Q/A = 300/(5)4 = 15$ ft/s

Then you can determine the water elevation at point B by the following:

$$d_2 = -\frac{1}{2}d_1 + \sqrt{\frac{2v_1^2 d_1}{g} + \frac{d_1^2}{4}} = -\frac{1}{2}(4) + \sqrt{\frac{2(15)^2(4)}{32.2} + \frac{4^2}{4}} = 5.74'$$

Then you can determine the velocity at point B:

$$v_2^2 = \left(\frac{gd_1}{2d_2}\right)(d_1 + d_2) = \left(\frac{32.2(4)}{5.74(2)}\right)(5.74 + 4); \quad v_2 = 10.45 \; ft/s$$

The answer is **(B)**

Solution W12

For a confined aquifer use the following equation:

$$Q = \frac{2\pi KY(y_1 - y_2)}{\ln\frac{r_1}{r_2}}$$

It is important to remember that the heights y, are from the base of the aquifer to the water table elevation. Therefore the equation is:

$$Q = \frac{2\pi KY(y_1 - y_2)}{\ln\frac{r_1}{r_2}} = \frac{2\pi(20)(50)((50-20)-(50-15))}{\ln\frac{40}{75}} = 49977 \; gal/day$$

The answer is **(D)**

Solution W13

First you can calculate the instantaneous demand at the ten-year period since it does not change per person using the multiplier:

$$Q_{Instant} = M(AADF) = 2.2(85) = 187 \; gpcd$$

Then we need to calculate the increased population at the ten-year period. Since the rate is constant, we can use the following equation:

$$P_{10} = P_o(1+i)^n = 25000(1+0.03)^{10} = 33598 \; people$$

Now to get the amount of water necessary:

V = 33598(187) = 6.28 x 10^6

The answer is **(D)**

Solution W14

To answer this question is to understand how WLA's are defined. WLA's are identified as a point source under the Clean Water Act. Therefore, the source must be an entity that can be identified. In the options presented the wastewater treatment plant would fall under this requirement. The rest conversely are nonpoint source allocations which generally occur from more widespread sources such as runoff or drainage.

The answer is **(B)**

Solution W15

The critical concept here is to understand that during these operations, the backwash water rate of rise must not exceed the settlement rate of the particles. Therefore, to design for the operation time we must set the rise rate equal to the settlement velocity of the sand. Since we are given Reynolds number and it is less than 1, we know we can use Stokes Law:

$$v_s = \frac{(SG_{particle}-1)D_{ft}^2 g}{18v} = \frac{(2.65-1)0.0002^2(32.2)}{18(9.62 \times 10^{-6})} = 0.0122 \; ft/sec = 0.74 \; ft/m$$

Then you can solve for the time:

$$V = A_{Filter}(Rise \; Rate)t_{Backwash} = 200'(10') = 200'(0.74 \; ft/m)t; \; t = 13.5 \; mins$$

The answer is **(B)**

Solution W16

It is first important to note that the lime is for the carbonate hardness removal and the soda ash for the noncarbonate. However, lime will first react with the CO_2 and therefore the amount of lime must account for both this and the carbonate hardness. First determine the carbonate hardness:

Carbonate Hardness = Total – Noncarbonate Hardness = 200-50 = 150 mg/L

Therefore, the amount of lime needed at 92% purity = $\frac{150+20}{0.92}$ = 184.78 mg/L as $CaCO_3$

The soda ash is then used to remove the noncarbonate at 95% = $\frac{50}{0.95}$ = 52.6 mg/L as $CaCO_3$

The answer is **(D)**

Solution W17

A trapezoidal weir with a 4:1 ratio of height to width of the sides has the following equation for discharge:

$$Q = 3.367bH^{3/2} = 3.367(1)0.5^{1.5} = 1.19 \frac{ft^3}{s}$$

The answer is **(A)**

Solution W18

The pond will overtop if the capacity is exceeded within the given time period of 6 hours. The amount that will overtop the pond is 200000 – 80000 = 120000 ft^3. Then we can get an average flow rate over the 6 hr period: 120000/(6.0(60)(60)) = 5.56 ft^3/s

Then we need to determine the intensity which will cause the three drainage areas to cause this flow rate. The flow into the pond is the sum of the three areas using the rational method:

$Q_T = A_1C_1i + A_2C_2i + A_3C_3i$

5.56 = i((5)(.16) + (4)(0.23) + (6)(0.22)); Solve for i = 1.83 in/hr

The answer is **(B)**

Solution W19

First since this channel will be designed for critical flow, we must find the critical depth of water:

$$d_c^3 = \frac{Q^2}{gw^2} = \frac{500^2}{32.2(10)^2}; \quad d_c = 4.26'$$

Next use the Chezy Manning equation to determine the slope. First calculate the area and hydraulic radius using the critical depth:

A = 10(4.26) = 42.6 ft²

$$R = \frac{A}{P} = \frac{42.6}{10+2(4.26)} = 2.3'$$

Q = (1.49/n)AR$^{2/3}$S$^{1/2}$ = 500 = (1.49/0.015)(42.6)(2.3)$^{2/3}$S$^{1/2}$; S = 0.0046

The answer is **(B)**

Solution W20

This example uses the Bernoulli conservation equation and the principles associated. The equation is as follows:

$E_t = E_{pr} + E_v + E_p = p/\rho + v^2/2g + z$

And as we know that the energy is conserved throughout the system, we know that the energy is equal at each point in question.:

$E_A = E_B = E_C$

If we equate points A to point C, we know that there is no pressure component at both points since it is open to the atmosphere. We also are told the velocity at A is zero and therefore we are left with:

$E_A = E_C = z_A = v_C^2/2g = 35 = v_C^2/2(32.2)$; v_C = 47.5 ft/s

Then we can use the conservation of flow to determine the velocity at point B:

$Q_C = Q_B = V_C A_C = V_B A_B = 47.5\pi(1)^2 = V_B(\pi(2)^2)$; V_B = 11.9 ft/s

Then since energy is conserved you can set point B and C equal:

$E_B = E_C = p_B/\rho + v^2/2g + z_B = 35 = p_B/62.4 + (11.9)^2/2(32.2) + 5; p_B = 1734$ psf;

The answer is **(A)**

Solution W21

The rational method can be used to determine the peak discharge:

$Q = CiA = 0.25(1.8)(15) = 6.75$ ft³/s

The unit hydrograph peak discharge is found by taking the peak discharge and dividing it by the average precipitation across the drainage area. First find the average precipitation in the proper units:

$$P_{avg} = V/A_d = \frac{2\ ac\text{-}ft\left(43560\frac{ft^2}{ac}\right)\left(12\frac{in}{ft}\right)}{15\ ac\left(43560\frac{ft^2}{ac}\right)} = 1.6\ in$$

Then find the unit hydrograph peak discharge:

$Q_U = 6.75/1.6 = 4.22$ ft³/s-in

The answer is **(C)**

Solution W22

This example requires the use of Monod's equation. First since you are given a change in concentration over time you can determine the rate of growth for the microorganisms:

r_g = Rate of growth = $\frac{dX}{dt} = \frac{13.5-9.0}{6.0-0} = 0.75 \frac{mg}{L}/hr$

Then you can determine the maximum specific growth rate coefficient by plugging in the appropriate variables to the Monod equation:

$$r_g = \frac{\mu_m X S}{K_s + S}$$

$$0.75 = \frac{\mu_m(13.5)(5.0)}{200 + 5.0}, \mu_m = 2.28\ hr^{-1}$$

The answer is **(D)**

Solution W23

While pipe networks are difficult to analyze, the principles remain and, in this case, we can use the conservation of mass. Not only can you analyze the system as a whole, but the flow into and out of each node must be equal. Therefore, the final diagram is as follows:

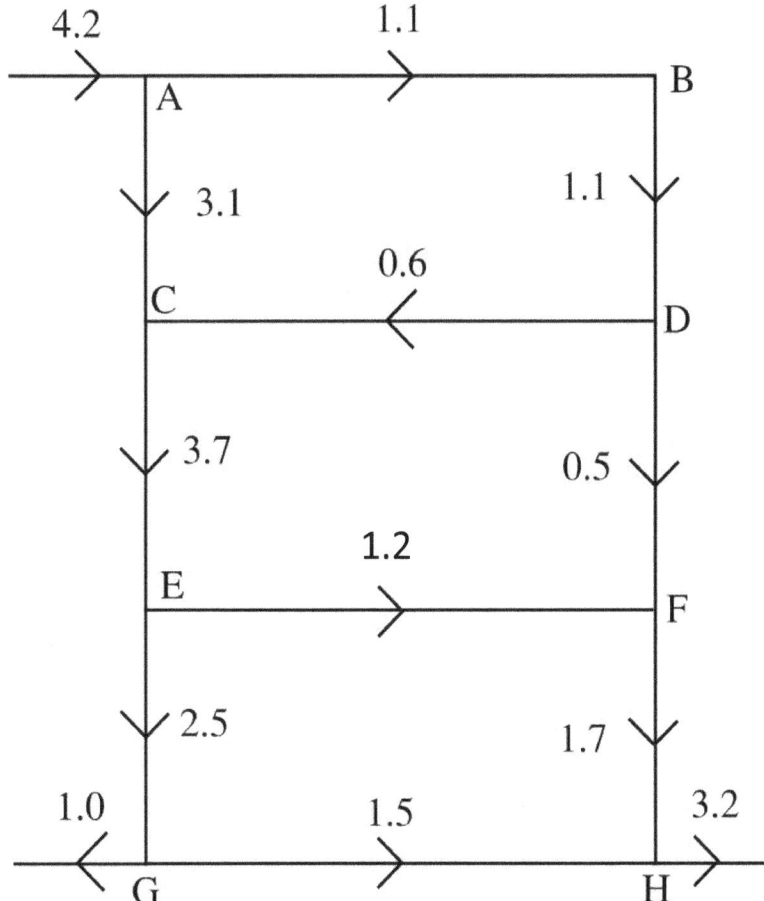

Solution W24

The horse power needed is a function of the total dynamic head. This is the addition of the static head (the change in elevation) and the head loss due to friction. First calculate the headloss due to friction from the Darcy Equation:

$$h_f = \frac{fLv^2}{2Dg} = \frac{(0.02)(40')\left(4/(\pi 0.25^2)\right)^2}{2(0.5)(32.2)} = 10.3'$$

Then add this to the static head: (175 – 150) + 10.3 = 35.3'

Then find the horsepower:

$$h_T = \frac{\left(550 \frac{ft-lb}{sec-hp}\right) P_{hp,input} \eta_{pump}}{Q\gamma} = 35.3 = \frac{\left(550 \frac{ft-lb}{sec-hp}\right) P(0.8)}{4(62.4)} \quad P = 20 \; hp$$

The answer is **(A)**

Solution W25

This can be solved by the dilution purification equation. However, the flow rate needs to be in the same units:

$$\frac{5.0 \; MGD \left(10^6 \frac{gal}{MG}\right)}{\left(7.48 \frac{gal}{ft^3}\right)\left(24 \frac{hr}{day}\right)\left(60 \frac{min}{hr}\right)\left(60 \frac{s}{min}\right)} = 7.73 cfs$$

$$C_f = \frac{C_1 Q_1 + C_2 Q_2}{Q_1 + Q_2} = 10.2 = \frac{C_1 3.2 + 7.73(8.0)}{3.2 + 7.73}; \; C_1 = 15.51 \; mg/L$$

The answer is **(D)**

Solution W26

Flow rate through an orifice is determined by the following equation:

$$C_f A_o \sqrt{\frac{2g_c(p_1 - p_2)}{\rho}}$$

First calculate the area of pipe and of the orifice:

$A_1 = \pi(4/12)^2 = 0.35 \text{ ft}^2$
$A_o = \pi(1/12)^2 = 0.022 \text{ ft}^2$

Then we must determine the velocity of approach factor which we can use for the flow coefficient:

$$C_f = C_d F_{va}$$

$$F_{va} = \frac{1}{\sqrt{1 - \left(\frac{C_c A_o}{A_1}\right)^2}} = \frac{1}{\sqrt{1 - \left(\frac{(0.64)(0.022)}{0.35}\right)^2}} = 1.0$$

$$C_f = C_d F_{va} = 0.6(1.0) = 0.6$$

Then calculate the flow rate:

$$Q = C_f A_o \sqrt{\frac{2g_c(p_1 - p_2)}{\rho}} = 0.6(0.022)\sqrt{\frac{2(32.2)(1500)}{62.4}} = 0.52 \text{ } ft^3/s$$

The answer is **(A)**

Solution W27

The type of hardness in a sample can be characterized by comparing it to the alkalinity of a sample. So we must determine the alkalinity but as mg/L of $CaCO_3$. To do this the ions have factors which convert them. Therefore, the alkalinity is:

0.82(30) + 1.67(20) = 58 mg/L as $CaCO_3$

Then since 58 > 50, the alkalinity is greater than the hardness. From this we can determine that all hardness is then carbonate.

The answer is **(B)**

Solution W28

The discharge for a submerged culvert can be determined by the following equation:

$$Q = C_d A \sqrt{2gh_{eff}}$$

$$h_{eff} = h - h_{f,culvert} - h_{f,entrance}$$

The effective head loss is the difference in elevation minus the head loss of the pipe and the minor loss at the entrance. The equation can be rewritten as:

$$h_{eff} = h - \frac{fLv^2}{2D} - K\frac{v^2}{2g} = 5' - \frac{(0.09)(50')(6)^2}{2(5)(32.2)} - 0.5\frac{6^2}{2(32.2)} = 5' - 0.5' - 0.28' = 4.22'$$

Then calculate the discharge:

$$Q = C_d A \sqrt{2gh_{eff}} = 0.98(\pi 2.5^2)\sqrt{2(32.2)(4.22)} = 317 \; ft^3/s$$

The answer is **(C)**

Solution W29

Chlorine when mixed with water forms hydrochloric and hypochlorous acids. However, when the PH is greater than 9, the hypochlorous acid also becomes H⁺ + OCl⁻. O_2 is not formed from these reactions.

The answer is **(B)**

Solution W30

First determine the flow rate from the spillway:

$$Q = C_s b(H)^{3/2} = (3.80)(15)(5)^{\frac{3}{2}} = 637.3 \text{ cfs}$$

The water from the spillway hits the channel and then has to stabilize causing a hydraulic jump. Therefore, we can use the following equation to determine the height downstream:

$$d_2 = -\frac{1}{2}d_1 + \sqrt{\frac{2v_1^2 d_1}{g} + \frac{d_1^2}{4}}$$

First calculate the velocity from the flow rate:

v = 637.3/((20)(2)) = 15.9 ft/s

Then determine the height from the jump:

$$= -\frac{1}{2}(2) + \sqrt{\frac{2(15.9)^2(2)}{32.2} + \frac{2^2}{4}} = 4.7'$$

The answer is **(D)**

Solution W31

Cavitation is the vaporization of water in a pump which causes bubbles to form and decrease pump performance. Increasing pump speeds will cause more air flow in the fluid and therefore increases the chance for cavitation.

The answer is **(B)**

Solution W32

The discharge of the channel can be approximated by taking average depths and velocities along the width:

$$0\text{-}1 = w\left(\frac{y_0+y_1}{2}\right)\left(\frac{v_0+v_1}{2}\right) = 4\left(\frac{0+3}{2}\right)\left(\frac{0+0.3}{2}\right) = 0.9 \text{ cfs}$$

$$1\text{-}2 = 4\left(\frac{3+6}{2}\right)\left(\frac{0.3+0.6}{2}\right) = 8.1 \text{ cfs}$$

$$2\text{-}3 = 4\left(\frac{6+10}{2}\right)\left(\frac{0.6+1.2}{2}\right) = 28.8 \text{ cfs}$$

$$3\text{-}4 = 4\left(\frac{10+5}{2}\right)\left(\frac{1.2+0.8}{2}\right) = 30 \text{ cfs}$$

$$4\text{-}5 = 4\left(\frac{5+4}{2}\right)\left(\frac{0.8+0.4}{2}\right) = 10.8 \text{ cfs}$$

$$5\text{-}6 = 4\left(\frac{4+0}{2}\right)\left(\frac{0.4+0}{2}\right) = 1.6 \text{ cfs}$$

Total discharge = Σq = 0.9 + 8.1 + 28.8 + 30 + 10.8 + 1.6 = 80.2 cfs

The answer is **(B)**

Solution W33

The total time of concentration is the sum of the two components. The first being sheet flow from the drainage area and the second shallow flow from the swale. You can determine the time of concentration for shallow flow by the following:

$$t_c = t_{sheet} + t_{shallow} = 35 = 30 + t_s; \ t_s = 5 \ min$$

The slope is related to the velocity by the following equation for unpaved sections:

$$v_{shallow, ft/s} = 16.1345\sqrt{S_{Decimal}} = 16.1345\sqrt{0.02} = 2.28 \ ft/s$$

Then the travel time is equal to the length over the velocity. Therefore:

L = vt_s = (2.28 ft/s)(5 min)(60 s/min) = 684'

Solution W34

This question is simply an application of the Horton infiltration equation:

$$f_p = f_c + (f_o - f_c)e^{-kt}$$

f_p = Infiltration rate at time t (in/hr)
f_c = Ultimate infiltration rate (in/hr)
f_o = Initial infiltration rate (in/hr)
k = Decay constant (time^{-1})
t = Time (hr)

Plugging in yields:

$$f_p = 2 + (10 - 2)e^{-(2)(25/60)} = 5.48 \text{ in/hr}$$

The answer is **(B)**

Solution W35

The flow rate for a gutter can be determined from the following:

$$Q = K(^Z/_n)s^{1/2}y^{8/3}$$

Plugging in the variables provides:

$$Q = 0.56 \left(\frac{(^1/_{0.025})}{0.012} \right) 0.003^{1/2} 0.25^{8/3} = 2.53 \; cfs$$

Remember that K is taken as a constant = 0.56 ft³/(s-ft)

The answer is **(A)**

Solution W36

The flow rate can be determined by the peak flow plus any infiltration and inflow. The peak flow rate is determined by $Q_{Avg}*PF$. The peak factor can be approximated by the Harmon equation:

$$PF = \frac{18 + \sqrt{P}}{4 + \sqrt{P}} = \frac{18 + \sqrt{36}}{4 + \sqrt{36}} = 2.4$$

Then determine the peak flow:

Q_{Peak} = 2.4(3.6) = 8.64 MGD

Then increase for infiltration and inflow:

Q = 8.64(1.05) = 9.1 MGD

The answer is **(D)**

Solution W37

This question relies on the volume balance for a lake. Since transpiration and water release are ignored, the following equation applies:

$$\Delta S = P + R + GI - GO - E$$

The precipitation is provided in inches/hr and needs to be converted to a volume based on the approximation of the surface area = $\pi r^2 = \pi(60')^2 = 11309.7$ ft^2

To get the volume of precipitation = (0.4/12)(10 hrs)(11309.7) = 3770 ft^3 (7.48 gal/ft^3) = 28199 gal

Then the equation becomes:

521650 − 500000 = 28200 + 12000 − 4250 − E; E = 14300 gallons = 1911.76 ft^3

Then convert to in/hr

E = (1911.76/(11309(12 hrs)))(12 in/ft) = 0.17 in/hr

The answer is **(C)**

Solution W38

Coagulation and Sedimentation and Rapid Sand Filtration is needed when the suspended solids are above 100 mg/L.

Lime Softening is needed when the CaCO₃ concentration is greater than 200 mg/L.

Postchlorination is related to the treatment of coliform which is not present.

Salt Water Conversion is only needed if the chlorides are above 500 mg/L, here we have 100.

Aeration is needed with the iron and manganese levels above 1.0.

The answers are **(A), (B), (D), and (F)**

Solution W39

First you can determine the Soil water storage, S, from the curve number:

$$S = \frac{1000}{CN} - 10 = \frac{1000}{49} - 10 = 10.4 \ in$$

The equation for peak discharge from a synthetic hydrograph is:

$$Q_p = \frac{0.756 A_{d,Acres}}{t_p}$$

However, the time to peak flow must first be calculated:

$$t_p = 0.5 t_R + t_1 \ where \ t_1 = \frac{L_o^{0.8}(S+1)^{0.7}}{1900\sqrt{S_{percentage}}} = \frac{4400^{0.8}(10.4+1)^{0.7}}{1900\sqrt{2}} = 1.68 \ hrs$$

$$t_p = 0.5(6) + 1.68 = 4.68 \ hrs$$

$$Q_p = \frac{0.756(2050)}{4.68} = 328.5 \ cfs$$

The answer is **(A)**

Solution W40

The capitalized cost is the amount of money needed to support yearly costs on interest alone for infinite life. You can determine capitalized cost by the following equation:

Capitalized cost = initial cost + yearly cost/interest rate.

In this case the initial cost is zero and the capitalized cost is:

$$C = \frac{(10000 + 6000 + 4000)}{0.08} = \$250,000$$

The answer is **(C)**

Answer Key

M1	A	M41	C	W1	B
M2	C	M42	D	W2	A
M3	ACDGH	M43	326.4	W3	C
M4	D	M44	D	W4	A
M5	A	M45	A	W5	A
M6	B	M46	B	W6	C
M7	See Solution	M47	B	W7	A
M8	A	M48	B	W8	A
M9	C	M49	C	W9	See Solution
M10	D	M50	C	W10	A
M11	B	M51	A	W11	B
M12	B	M52	C	W12	D
M13	10+60	M53	B	W13	D
M14	A	M54	D	W14	B
M15	D	M55	D,E	W15	B
M16	A	M56	See Solution	W16	D
M17	C	M57	B	W17	A
M18	D	M58	D	W18	B
M19	B	M59	B	W19	B
M20	C	M60	B	W20	A
M21	C	M61	C	W21	C
M22	A	M62	A	W22	D
M23	D	M63	A	W23	See Solution
M24	C	M64	D	W24	A
M25	A	M65	A	W25	D
M26	See Solution	M66	C	W26	A
M27	B	M67	A	W27	B
M28	C	M68	A	W28	C
M29	A	M69	D	W29	B
M30	A	M70	B	W30	D
M31	B	M71	C	W31	B
M32	D	M72	D	W32	B
M33	D	M73	B	W33	684'
M34	C	M74	C	W34	B
M35	A	M75	A	W35	A
M36	C	M76	C	W36	D
M37	C	M77	A,B,D,F	W37	C
M38	C	M78	D	W38	A,B,D,F
M39	B	M79	C	W39	A
M40	B	M80	C	W40	C

Thank You Again for Your Purchase!

What Did You Think of the Practice Exams?

It is a long and difficult road to passing and we are extremely grateful you chose us to help along the way. We hope that it added value and efficiency to your studying. We are here for any questions or concerns you may have and we will respond quickly if you email us at:

PECoreConcepts@gmail.com

If you enjoyed this book, it would help greatly if you have the time to **leave a positive review on our Amazon product page**. Reviews help to support small businesses like ours.

References

Civil Engineering Reference Manual for the PE Exam Fifteenth edition 2015. Michael R. Lindeburg Professional Publications Inc. (PPI)

Water and Wastewater Calculations Manual Second Edition 2007. Shun Dar Lin. The McGraw-Hill Companies inc.